Safe Water
From Every Tap

Improving Water Service
to Small Communities

Committee on Small Water Supply Systems

Water Science and Technology Board

Commission on Geosciences, Environment, and Resources

National Research Council

NATIONAL ACADEMY PRESS
Washington, D.C.   1997

NATIONAL ACADEMY PRESS • 2101 Constitution Avenue, NW • Washington, DC 20418

NOTICE: The project that is the subject of this report was approved by the Governing Board of the National Research Council, whose members are drawn from the councils of the National Academy of Sciences, the National Academy of Engineering, and the Institute of Medicine. The members of the committee responsible for the report were chosen for their special competences and with regard for appropriate balance.

This report has been reviewed by a group other than the authors according to procedures approved by a Report Review Committee consisting of members of the National Academy of Sciences, the National Academy of Engineering, and the Institute of Medicine.

Support for this project was provided by the U.S. Environmental Protection Agency.

Library of Congress Catalog Card Number 96-70443
International Standard Book Number 0-309-05527-X

Safe Water From Every Tap: Improving Water Service to Small Communities is available from the National Academy Press, 2101 Constitution Avenue, NW, Lockbox 285, Washington, DC 20055 (1-800-624-6242; http://www.nap.edu).

Cover art by Y. David Chung. Chung is a graduate of the Corcoran School of Art in Washington, D.C. He has exhibited his work throughout the country, including the Whitney Museum in New York, the Washington Project for the Arts in Washington, D.C., and the Williams College Museum of Art in Williamstown, Massachusetts.

Printed in the United States of America

TD
223
523
1997

COMMITTEE ON SMALL WATER SUPPLY SYSTEMS

The National Academy of Sciences is a private, nonprofit, self-perpetuating society of distinguished scholars engaged in scientific and engineering research, dedicated to the furtherance of science and technology and to their use for the general welfare. Upon the authority of the charter granted to it by the Congress in 1863, the Academy has a mandate that requires it to advise the federal government on scientific and technical matters. Dr. Bruce Alberts is president of the National Academy of Sciences.

The National Academy of Engineering was established in 1964, under the charter of the National Academy of Sciences, as a parallel organization of outstanding engineers. It is autonomous in its administration and in the selection of its members, sharing with the National Academy of Sciences the responsibility for advising the federal government. The National Academy of Engineering also sponsors engineering programs aimed at meeting national needs, encourages education and research, and recognizes the superior achievements of engineers. Dr. William A. Wulf is interim president of the National Academy of Engineering.

The Institute of Medicine was established in 1970 by the National Academy of Sciences to secure the services of eminent members of appropriate professions in the examination of policy matters pertaining to the health of the public. The Institute acts under the responsibility given to the National Academy of Sciences by its congressional charter to be an adviser to the federal government and, upon its own initiative, to identify issues of medical care, research, and education. Dr. Kenneth I. Shine is president of the Institute of Medicine.

The National Research Council was organized by the National Academy of Sciences in 1916 to associate the broad community of science and technology with the Academy's purposes of furthering knowledge and advising the federal government. Functioning in accordance with general policies determined by the Academy, the Council has become the principal operating agency of both the National Academy of Sciences and the National Academy of Engineering in providing services to the government, the public, and the scientific and engineering communities. The Council is administered jointly by both Academies and the Institute of Medicine. Dr. Bruce Alberts and Dr. William A. Wulf are chairman and interim vice chairman, respectively, of the National Research Council.

Preface

Small water supply systems are an important part of the drinking water industry in the United States. Approximately 20 percent of the U.S. population is served by more than 54,000 systems, each serving 10,000 or fewer people, and approximately two-thirds of these systems serve communities with populations of 500 or fewer people. The numbers of such systems are increasing rapidly; for example, the number of water systems serving 500 or fewer people increased sevenfold, from 5,000 to more than 35,000, between 1963 and 1993. While many of these systems produce a safe, wholesome water supply, many others lack the capital needed to upgrade their facilities and the revenue needed for day-to-day operation and maintenance. Modification of these systems to meet new standards, and implementation of monitoring programs to ensure that water quality is adequate, are particularly troublesome for such communities. The problems are compounded by the fact that small water systems lack the economies of scale of larger systems. These problems are most acute for systems serving fewer than 500 people. These systems violate drinking water standards for microbes and chemicals more than twice as often as systems serving more than 10,000 people. Communities with 500 or fewer residents are thus more vulnerable than larger communities to outbreaks of waterborne disease.

The U.S. Environmental Protection Agency (EPA) recognizes the problems of small water supply systems and in 1994 asked the National Research Council (NRC) to study the problem. In response to this request, and with the sponsorship of the EPA, the NRC's Committee on Small Water Supply Systems was established. Its membership consisted of 12 experts in water treatment, utility management, finance, and public health. In this report, the committee proposes a solution to the problem of providing safe drinking water to small communities that has three elements, each of which is equally important:

1. providing affordable water treatment technologies,
2. creating the institutional structure necessary to ensure the financial stability of the water systems, and
3. improving programs to train small system operators in all aspects of water system maintenance and management.

The committee studied the problems of small water systems by inviting a large number of water industry representatives to address the committee in their areas of expertise. These included representatives from the EPA who were involved in developing regulations for small water supplies and in researching water treatment processes for small water supplies; representatives of state regulatory agencies, with emphasis on states that have implemented innovative approaches to managing small systems; manufacturers of equipment for small systems; a representative from a third-party testing organization that evaluates small systems equipment and point-of-entry/point-of-use devices; a representative from the Natural Resources Defense Council; and representatives of groups responsible for providing assistance to small communities. The report is based on a thorough review of the information presented by these individuals, information from the published literature, and the expertise of the committee members.

The successful preparation of this report was in large part dependent on the skills of Jackie MacDonald, NRC senior staff officer, who pressured and cajoled us into getting our sections of the report done in a timely fashion, contributed original written material for significant sections of the report, and thoroughly edited the entire report. Jackie's attention to detail, persistence, and organization were essential to the timely completion of this report. Jackie was ably assisted by Etan Gummerman, research associate, and Anita Hall, administrative assistant. David Dobbs, editor, made significant contributions to improving the clarity of several of the report chapters. Their important input is gratefully acknowledged. Also essential to the completion of this project was Stephen Clark, the committee's liaison from the EPA. His valuable insights and responsiveness to all of the committee's requests for information greatly facilitated the committee's work.

The efforts of the committee members in attending meetings, researching their subjects, writing and revising their contributions, and reviewing and revising the entire report are acknowledged and sincerely appreciated. I hope that the reader will agree with me that the committee has done its job very well.

VERNON L. SNOEYINK, *Chair*
Committee on Small Water Supply Systems

Contents

Executive Summary

U.S. citizens generally expect to be able to drink their tap water with minimal health risk. While the quality of U.S. drinking water is superior to that in many parts of the world, not all U.S. citizens are receiving the same quality of water service. For example, during one recent 27-month period, 23.5 percent of U.S. community water systems violated safe drinking water standards one or more times for microbes that indicate the possible presence of bacteria, viruses, or parasites associated with human illnesses. Nearly 600 waterborne disease outbreaks have been reported in the past two decades.

Meeting drinking water standards is most difficult for water systems in small communities. Small communities often cannot afford the equipment and qualified operators necessary to ensure compliance with safe drinking water standards. Increases in both the number of drinking water regulations and the number of small community water systems over the past three decades have compounded the problem of providing safe drinking water to small communities. For example, the number of water systems serving 500 or fewer people increased sevenfold, from 5,000 to more than 35,000, between 1963 and 1993; the number of systems serving 501 to 10,000 people increased by more than 60 percent. Over this same time period, the number of contaminants regulated by federal drinking water standards increased from fewer than 20 to more than 100.

This report focuses on how to provide safe drinking water to small communities. It discusses technologies for small water systems, how to streamline pilot testing of these technologies to make them more affordable, financing and management of small systems to ensure their sustainability, and training of small

1

system operators. The report was written by the National Research Council's Committee on Small Water Supply Systems. The committee was appointed in 1994 at the request of the U.S. Environmental Protection Agency (EPA) to study the problem of providing water service to small communities. Its membership consisted of 12 experts in water treatment, utility management, finance, and public health.

As discussed in this report, the solution to the problem of providing safe drinking water to small communities has three elements, each equally important: (1) providing affordable water treatment technologies, (2) creating the institutional structure necessary to ensure the financial stability of water systems, and (3) improving programs to train small system operators in all aspects of water system maintenance and management.

STATUS OF SMALL SYSTEMS

More than 54,000 small water systems (defined for this report as those serving 10,000 or fewer people) provide drinking water to approximately 20 percent of the U.S. population. Sixty-six percent of these systems serve communities with populations of 500 or fewer.

While some small communities are in wealthy areas, most small communities have difficulty raising the capital needed to upgrade their water systems and the revenue needed for day-to-day water system operation and maintenance. In extreme cases, these small communities can lack water service altogether. For example, as of 1990, more than 1.1 million U.S. households lacked plumbing.

Capital and adequate operating revenue are most difficult to obtain for small communities in nonmetropolitan areas. Average incomes in the smallest of these communities are one-third lower than incomes in larger, metropolitan areas. Unemployment rates can be more than 50 percent higher than those in metropolitan areas. Lenders are often unwilling to provide loans to rural communities because of the small profits generated by these loans. Whether a small system is located in a rural area or a metropolitan one, it will lack the economies of scale of larger communities in providing water service; per-person costs for water service must be higher in small communities than in larger ones to provide the same level of service because the costs are spread over a smaller population.

Small communities that lack adequate revenue for water treatment and distribution can have difficulty complying with the Safe Drinking Water Act. For example, systems serving fewer than 500 people violate drinking water standards for microbes and chemicals more than twice as often as those serving larger communities. Such violations leave these communities vulnerable to outbreaks of waterborne illness. In addition, the large number of violations in small communities poses a serious management problem for the state regulatory agencies responsible for implementing the Safe Drinking Water Act.

EVALUATING TECHNOLOGIES FOR SMALL SYSTEMS

Before looking to technological answers to water quality problems, small water supply systems should exhaust other available alternatives for improving water quality. One option is to find a higher-quality source water, such as by switching from surface water to ground water or relocating a well to a cleaner aquifer. In general, ground water sources are a better choice for small water systems than surface water sources because they are less turbid and have lower concentrations of microbiological contaminants than surface water. A second, nontechnical option for improving small system water quality is to purchase treated water from a nearby utility. Such options are often more cost effective than attempting to remove contaminants from a poor-quality source water.

When other options are not available and small systems must turn to water treatment processes in order to provide water that meets the requirements of the Safe Drinking Water Act, they may have difficulty raising revenue for capital improvements. One option available for reducing the costs of water treatment for these communities is the use of preengineered "package plants." Package plants are off-the-shelf units that group elements of the treatment process, such as chemical feeders, mixers, flocculators, sedimentation basins, and filters, in a compact assembly. Package plants do not eliminate the need for an engineer to design the specifics of the on-site application of water treatment equipment. Nevertheless, because package systems use standard designs and factory-built treatment units that are sized, assembled, and delivered to the customer instead of being custom built on site, such systems have the potential to significantly reduce the engineering and construction costs associated with a new water treatment system.

Site-specific pilot testing requirements can significantly increase the costs of package water treatment plants, partially offsetting the cost savings these systems offer. State regulators often require pilot tests of all new treatment systems other than chlorinators. Often package plants must be evaluated over and over again for source waters having similar quality but located in different communities. Pilot tests can last anywhere from several weeks to 1 year or more. Extensive pilot testing reduces the savings achieved by having the package plants designed and assembled at a central facility. Manufacturers have reported that pilot testing can increase the costs of their equipment by more than 30 percent. For example, according to one manufacturer, a 6-month pilot test can add $16,000 to the cost of a $45,000 package filtration system.

Certification of package plant performance by an independent third party would reduce package plant costs by reducing, although not eliminating, the need for site-specific testing. Currently, no national program exists for certifying drinking water treatment systems other than point-of-use (POU) and point-of-entry (POE) devices used in individual homes. The National Sanitation Foundation (NSF) International, which certifies in-home water treatment equipment, is currently cooperating with the EPA to develop a verification program for package plants. This program, launched in late 1995, is in its beginning phases and is currently funded for a 3-year period. Support for the program should continue, because it could reduce the costs of drinking water treatment technologies for small communities. Once the program is established, testing fees provided by equipment manufacturers will sustain most of its costs.

A key component of a national pilot testing and verification program for package plants is standard protocols for equipment testing. Currently, such protocols do not exist. Water treatment system designers generally conduct bench and pilot studies using their own individual methods and parameters for documenting water quality. As a consequence, it is difficult to compare data sets developed by different investigators. Establishment of standard protocols that measure the parameters covered in Safe Drinking Water Act regulations would allow data collected in one location to be applied elsewhere.

Another key component of a package plant testing and certification program is a national data base for reporting test results. Currently, no such data base exists. Considerable "reinvention of the wheel" occurs as new tests are required to verify technologies at each new location even if identical tests were performed elsewhere on water of a similar quality. Such a data base could be created by expanding the Registry of Equipment Suppliers of Treatment Technologies for Small Systems (RESULTS) data base at the National Drinking Water Clearinghouse in West Virginia. The expanded data base should cover all of the available technologies, use standard formats for reporting data, and include complete information about raw water quality, finished water quality, and operation and maintenance costs for each technology.

While development of standard protocols for testing drinking water treatment technologies is a desirable goal, it is essential to recognize that the degree to which pilot testing can be centralized in order to reduce site-specific testing varies considerably depending on the type of technology, the nature of the water to be treated, and the availability of data documenting the performance of the technology on waters of various qualities. For many technologies, some aspects of process performance can be tested in a central facility, while others need to be evaluated for each source water treated. Following are some general principles that apply to pilot testing of various classes of water treatment processes (see Chapter 4 for details):

• Site-specific pilot testing of *aeration systems* is not necessary; performance can be predicted with design equations.
• For *membrane systems*, much of the detailed evaluation can be based on pilot tests or full-scale applications elsewhere. However, systems using ground water will need to evaluate the potential for chemical scaling of the membranes. Surface water systems will need to test the potential for the source water to foul

the membranes and determine whether pretreatment is required to remove particulate matter ahead of the membranes.

• For *granular activated carbon adsorption systems*, some degree of source water-specific testing is necessary because the ability of the carbon to adsorb a target contaminant varies significantly with the chemical composition of the raw water. In cases where the raw water has a low concentration of organic matter, such as in ground water, inexpensive bench-scale columns can adequately predict performance; for surface water systems, pilot tests will be necessary.

• *Powdered activated carbon adsorption systems* need to be evaluated in bench-scale tests, at a minimum, to determine the effectiveness of the powdered activated carbon on the particular raw water and with the mixing characteristics present in the system.

• *Ion exchange* and *activated alumina* systems require some degree of source water-specific bench- or pilot-scale evaluation to determine the potential for competitive adsorption of ions other than the target contaminants, which can affect the life of materials used in treatment.

• Because of the complexity of the chemical processes involved, *coagulation/filtration systems* require site-specific testing unless an identical coagulation/filtration system is already being used successfully on the same source water. The degree of testing required depends in part on the design of the system and in part on the characteristics of the raw water. In some cases, bench-scale tests using jars to determine appropriate coagulant doses will be adequate.

• *Diatomaceous earth filtration systems* require a few weeks of pilot testing to establish the effectiveness of different grades of diatomaceous earth and to estimate the length of filter runs that might be expected with a full-scale plant.

• For *slow sand filtration systems*, site-specific pilot testing is necessary, unless a slow sand filter is already treating the same source water at another location, because understanding of these systems is insufficient to allow engineers to predict what filtered water turbidity an operating slow sand filter will attain. Piloting of these systems need not be expensive. Pilot test units can be constructed from manhole segments and other prefabricated cylindrical products.

• *Bag filters* and *cartridge filters* need not be pilot tested at each site. Performance of these filters depends on careful manufacture of the equipment and its use on waters of appropriate quality rather than on manipulation of the water or equipment during treatment.

• *Lime softening systems* need not be pilot tested for small systems using ground water sources; jar testing to determine appropriate process pH and chemical doses is sufficient. Lime softening systems do need to be pilot tested if used on surface water sources with variable quality.

• *Disinfection systems* using free chlorine, chloramine, chlorine dioxide, or ozone need not be tested at each individual site. The effectiveness of these systems is predicted based on laboratory test results, which regulators consider to be applicable to all systems.

• Current regulations allow small systems to base *corrosion control* strategies on desk-top reviews of water quality, rather than on pilot tests.

For the smallest of water systems, in particular those serving a few dozen homes or less, POE or POU water treatment systems may provide a low-cost alternative to centralized water treatment. In POE systems, rather than treating all water at a central facility, treatment units are installed at the entry point to individual households or buildings. POU systems treat only the water at an individual tap. If a source water has acceptable quality for drinking except for exceeding the nitrate or fluoride standards, for example, using a POU system to treat the small number of liters per day needed for drinking and cooking might be less costly than installing a central treatment system that could remove the nitrate or fluoride from all water used by the community. Similarly, POE systems can save the cost of installing expensive new equipment in a central water treatment facility. POU and POE systems can also save the considerable costs of installing and maintaining water distribution mains when they are used in communities where homeowners have individual wells.

Regulators often have significant objections to using POE and POU devices. Concerns include the potential health risk posed by not treating all the water in the household (a problem for POU systems), the difficulty and cost of overseeing system operation and maintenance when treatment is not centralized, and liability associated with entering customers' homes. These objections have merit, particularly as system size increases and the complexity of monitoring and servicing the devices increases. Using centralized water treatment should be the preferred option for very small systems, and POE or POU treatment should be considered only if centralized treatment is not possible.

Recommendations: Technologies for Small Systems

• Application of technology (other than disinfection) to improve water quality should be considered only after other options, such as finding a cleaner source of water or purchasing water from a nearby utility, have been exhausted.

• The EPA should continue support for the fledgling water treatment technology verification program that it recently initiated with the National Sanitation Foundation.

• The EPA should oversee development of standard protocols and reporting formats for pilot testing water treatment technologies, especially package plants.

• The EPA should establish a standard national data base for water treatment technology information by expanding the RESULTS data base at the National Drinking Water Clearinghouse. The data base should include complete information on source and finished water quality, in standard units, and costs for each technology. It should be made available electronically, via the Internet.

• State agencies responsible for regulating water systems should assign a staff member to continually evaluate the status of knowledge relating to the performance of various water treatment processes of potential use in their jurisdictions. As more performance information is generated on waters of similar quality, the extent of preinstallation testing can be reduced, thus reducing the costs to the small system.

ENSURING SMALL WATER SYSTEM SUSTAINABILITY

Affordable technologies can help small communities provide better quality water, but technologies alone will not solve the problems of small water supply systems. Without adequate management and revenues, small communities will be unable to maintain even low-cost technologies. Many small communities lack a fee structure that is adequate to generate the necessary operating revenues, let alone funds for capital improvements. In other communities, the population is too small and average incomes are too low to provide sufficient revenue no matter what the fee structure. Lack of revenue leads to a vicious circle: without funding, water systems cannot afford to hire good managers, but without good managers, water systems will have trouble developing a plan to increase revenues. Institutional changes are needed to decrease the number of unsustainable water systems—that is, the number of systems lacking the resources needed to meet performance requirements over the long term.

Like businesses, small water systems are experiencing greater external pressure to change in response to the increasing number of regulations and increasing customer expectations. Unlike businesses, however, small systems have generally not done effective business planning. States should encourage small systems to do such planning by developing formal public health performance appraisal programs. Such programs should require each regulated water system in the state to assess its short- and long-term ability to provide adequate quantities of water that meets Safe Drinking Water Act standards. States should provide operating permits only to water utilities that have satisfactorily completed a performance appraisal. Where performance appraisals reveal problems, the states should assist the small water system in resolving the problems.

Performance appraisals should include analyses of the following types of information:

- existence of health orders (for example, boil water orders) issued to the water system or waterborne disease outbreaks in the community;
- the system's record of response to these orders and outbreaks;
- violations of water quality standards, including monitoring requirements;
- the water system's methods for keeping track of its compliance with Safe Drinking Water Act standards;
- the number of staff and their levels of training;

- responses to sanitary surveys (on-site visits by state regulators to inspect system source water, facilities, and operations); and
- whether the water system has an adequate plan specifying how it will meet present and future demands at an affordable cost while complying with the Safe Drinking Water Act and other regulations.

While regulators have long considered waterborne disease outbreaks, compliance with drinking water standards, operator certification, and sanitary surveys when evaluating small water systems, the importance of a comprehensive, forward-looking plan has often been overlooked. Proper planning and financing are key elements in ensuring the sustainability of water systems. Developing a water system plan will cause the utility to examine itself closely and develop a road map for the future. The plan should include information on future trends in service area, population, land use policies, and water demands on both a short-term (next 5 years) and long-term (next 20 years) basis. Based on this demographic information, it should evaluate needed system improvements, the current budget, the expected future budget, and projected future rates necessary to sustain the budget. The level of detail in such plans will vary with the size of the system, with very small systems requiring less detailed plans than larger systems.

If the performance appraisal uncovers problems that compromise the system's sustainability, then the water system either must improve service on its own or restructure by delegating some or all of its responsibilities to another entity, such as a rural electric utility, regional water authority, local government, or investor-owned utility. Restructuring arrangements generally fit one of four categories:

1. *direct ownership*, in which a small system reaches an agreement with another authority to take over system ownership or joins with other nearby systems to form a regional agency;

2. *receivership or regulatory takeover*, in which the state takes responsibility for transferring management of a failing water system to another authority in cases where the system owner does not voluntarily relinquish control;

3. *contract service*, in which a contractor provides specific services, such as operation and maintenance, water quality monitoring, emergency assistance, and billing, on a routine basis; and

4. *support assistance*, in which another utility provides support such as training the small system operator to repair a chlorinator, helping the small system develop a financial management plan, sharing water storage facilities, or making joint purchases of supplies or water to get volume discounts.

Each of these options consolidates some portion of the management and operation of several water systems within a larger agency, reducing costs to the consumer. For example, restructuring may mean that the community no longer needs to pay for a qualified full-time water system operator if, through restructuring, several systems can share an operator.

While restructuring can reduce the costs of providing water service to small communities, several barriers can stand in the way of restructuring. Organizations may be unwilling to take over deteriorated systems if they fear being responsible for financing all the necessary system improvements. Similarly, they may fear being held liable if the troubled system is in violation of the Safe Drinking Water Act. In other cases, small system owners may be unwilling to relinquish control to another authority. Incentives need to be provided to encourage qualified organizations to take over management of unsustainable small water systems and to encourage small systems to enter into such arrangements.

Recommendations: Small Water System Sustainability

• States should establish programs requiring all water systems to conduct public health performance appraisals. Only systems that have successfully completed a performance appraisal should be issued an operating permit.

• The federal government should limit state revolving fund (SRF) monies for drinking water systems to states with official performance appraisal programs. This will ensure that federal funds are not used to prop up unsustainable systems.

• SRF monies should be made available to public- and investor-owned utilities for assisting in the restructuring of small water systems.

• Federal, state, and local governments should provide tax incentives to organizations that assume responsibility for failing small water systems (see Chapter 5 for details).

• State public utility commissions should allow adjustments to the rate base of larger utilities that assume responsibility for insolvent water systems so that the rate base and depreciation practices can reflect the costs of acquiring the failing system.

• The EPA should provide temporary waivers to utilities for liabilities associated with Safe Drinking Water Act violations in cases where the utility has acquired a failing water company. These waivers should be tied to reasonable compliance schedules.

TRAINING OPERATORS FOR SMALL SYSTEMS

Even a well-financed water system with the most advanced treatment technologies cannot deliver its water reliably unless its operators are trained adequately. While all 50 states have regulations for certifying water system operators, the programs for training these operators are disjointed and often fail to meet the needs of small system operators.

Training of small system operators is provided through a mix of state-run workshops, informal instruction from equipment vendors and state regulators, courses at technical schools or universities, American Water Works Association courses, and rural water associations. These programs are not coordinated in any way. In addition, most operator training programs (and state certification requirements) cover the general theories underlying operation of numerous types of water treatment processes, some of them quite advanced, while operators of smaller systems need specific, hands-on training in only the treatment technologies their systems use. Training and certification programs are particularly deficient in teaching operators about water system management and administration— two areas that are as essential to small water system operation as are treatment and distribution.

The Safe Drinking Water Act Amendments of 1986 authorized the EPA to spend up to $15 million per year to provide technical assistance to small communities struggling to comply with the act's requirements. While the EPA provides $6.5 million annually to the National Rural Water Association and the Rural Community Assistance Program for technical assistance to small water systems, this spending has not resulted in the types of coordinated training programs needed to ensure that all water system operators are adequately trained. More leadership is needed at the national level to improve training programs for small water system operators.

Recommendations: Operator Training

• Funds should be provided to the EPA to establish an organizational work unit, based at EPA headquarters, responsible for identifying the knowledge and skills necessary to operate all aspects of drinking water systems.

• The new EPA work unit should arrange for an independent organization, such as the National Training Coalition or the National Environmental Training Center for Small Communities, to develop multimedia tools to deliver the needed training to system owners and operators across the country.

• The operator training programs should cover all of the areas necessary for running a small water system, including metering, customer service, financing, administration, and human resources management, as well as water treatment, water distribution, and public health.

• The states or their agents, with EPA support and coordination, should deliver the training programs to operators.

• States should rewrite their operator certification laws for small systems to allow small system operators to be certified only for the treatment processes employed in their systems. At the same time, states should institute a requirement that operators have knowledge of all of the skill areas (metering, finance, and so on) necessary for small system management.

In summary, water service to many of the nation's small communities is currently inadequate. Improving the quality of water service to these communities will require a combination of approaches: finding high-quality water sources, streamlining pilot testing requirements to make technologies more affordable, creating incentives to consolidate the management and financial administration of small systems, and improving programs to train small water system operators. Any one of these approaches alone will be insufficient to solve the problems of small water systems. A water system lacking adequate revenues and a well-trained operator will be unable to afford or maintain equipment, no matter how inexpensive, for water supply, treatment, and distribution. Conversely, a water system with a well-trained operator and sound financial plan may be unable to meet drinking water standards unless it can obtain treatment systems that are within its budget. National and state leadership are needed to improve the delivery of quality water to small communities.

1

Small Water Supply Systems: An Unsolved Problem

In the United States today, most citizens expect to have access to low-cost, high-quality drinking water at their taps. While the U.S. drinking water supply is superior to water supplies in many parts of the world and reflects the high priority that Americans have placed on water quality, there is still room for significant improvement in water service. For example, 23.5 percent of all U.S. community water systems violated Safe Drinking Water Act microbiological standards one or more times between October 1992 and January 1995, and 1.3 percent violated chemical standards, according to data from the U.S. Environmental Protection Agency (EPA). Waterborne disease outbreaks still occur in the United States, providing a reminder that contaminated drinking water continues to pose health risks even in highly developed nations.

Small communities face the greatest difficulty in supplying water of adequate quality and quantity because they have small customer bases and therefore often lack the revenues needed to hire experienced managers and to maintain and upgrade their water supply facilities. Interruptions in water service due to inadequate management, as well as violations of drinking water standards, are problems for some of these systems. Although the problems of supplying drinking water through individually operated small water systems have long been known, the number of small water systems has continued to increase.

As part of its long-term effort to improve water supply service to small communities, the EPA asked the National Research Council (NRC) to recommend strategies for improving water service to the nation's small communities. The EPA asked that, as part of its analysis, the NRC examine whether streamlining of current pilot testing requirements is possible for "package" water treatment plants (off-the-shelf units like small versions of the custom-designed water treatment equipment used by larger utilities). This report addresses the EPA's request. It was prepared by the NRC's Committee on Small Water Supply Systems, appointed in 1994 in response to the EPA's request. The committee consisted of 12 experts in water treatment engineering, utility management and financing, environmental law, and public health. Members convened six times over an 18-month period to develop this report. The group received input from a wide range of stakeholders—including utility personnel, equipment manufacturers, state regulators, rural assistance program managers, and environmentalists— who are concerned about water supply service to small communities.

This chapter outlines the scope of the small systems problem and documents that although long known, the problems of small water systems have not been solved. Chapter 2 describes the status of small community water systems and the problems faced by consumers when these systems are unable to maintain adequate facilities. Chapters 3 and 4 respond to the EPA's request for guidance on testing package water treatment systems: Chapter 3 reviews the capabilities of various classes of technologies appropriate for small systems, and Chapter 4 advises on the degree to which testing of these technologies can be standardized. Chapter 5 reviews institutional options (including restructuring system management and developing sound financial plans) for improving water supply service to small communities. Chapter 6 recommends ways to improve the training of small water system operators.

INCREASING NUMBER OF SMALL SYSTEMS

For the purposes of this report, a public water system is considered small if it serves 10,000 or fewer people, although systems serving fewer than 500 people face the biggest challenges in providing adequate water service. The EPA divides public water systems into three categories: community water systems, which serve the same population all year; nontransient noncommunity water systems such as schools, factories, and hospitals with their own water supplies; and transient noncommunity water systems such as campgrounds, motels, and gas stations with their own water supplies (see Box 1-1 for the formal definitions). This report generally focuses on community water systems because the problems of serving an entire community are different from those associated with developing a water system for a single building or installation. For example, while community water systems must raise capital for improvements by issuing bonds, applying for grants or loans, or increasing their water rates, noncommunity systems typically finance improvements internally, through their overall budget for capital improvements (Task Force on Drinking Water Construction Funding and Regionalization, 1991). Nevertheless, much of the information in the report is also relevant to noncommunity water systems.

As shown in Table 1-1, the number of small community water systems has

BOX 1-1 Classes of U.S. Public Water Supply Systems

According to the EPA (1994), there are more than 190,000 public water systems in the United States (including those in U.S. territories and Native American lands). The EPA classifies any water distribution system as public if it supplies at least 15 service connections or at least 25 people for at least 60 days each year. The agency divides these public water systems into three categories:

• *Community water systems* provide drinking water to the same population all year. According to the EPA (1994), the 57,561 community systems in the United States as of 1993 served nearly 243 million people. (The remaining 15 million U.S. residents obtained their water from private wells or other systems serving fewer than 15 connections or 25 people.)

• *Nontransient noncommunity water systems* provide drinking water to at least 25 of the same people for at least 6 months each year. Schools, factories, and hospitals with their own water supplies are examples of such systems. The EPA (1994) estimated that as of 1993, 23,992 such systems served more than 6 million people.

• *Transient noncommunity water systems* provide drinking water to transitory populations in nonresidential areas. Examples are campgrounds, motels, and gas stations that have their own water supplies. According to the EPA (1994), 109,714 such systems existed by 1993 and served more than 15 million people.

increased substantially in the United States in the last three decades. For example, in 1963 there were approximately 16,700 water systems serving communities with populations of fewer than 10,000; by 1993 this number had more than tripled—to 54,200 such systems. Approximately 1,000 new small community water systems are formed each year (EPA, 1995).

The fragmented U.S. network for supplying drinking water contrasts significantly with water supply networks in many other parts of the developed world. In England and Wales, for example, 10 regional water organizations and 22 water companies provide water and sewerage service to 99 percent of the population of 50 million (Okun, 1977). The regional water agencies were the result of a national program initiated during World War II to streamline the allocation of water resources, in part to ensure that water supplies would be adequate for fighting fires caused by German bomb attacks. The program initially resulted in a reduction from 1,200 water systems serving 40 million people in 1945 to fewer than 200 systems serving 50 million people in 1972. The success of this program led to a much larger regionalization program in 1973, resulting in the creation of the 10 public water authorities, which were privatized in 1989. The initial consolidation of water systems was achieved by encouraging large cities to extend service to outlying small communities or, in areas with no large cities, by encouraging small communities to form regional water boards. In addition to water

TABLE 1-1 U.S. Community Water Systems: Size Distribution and Population Served

Population Served	Number of Community Systems Serving This Size Community[a]		Total Number of U.S. Residents Served by Systems This Size[b]	
	1963	1993	1963	1993
Under 500	5,433 (28%)	35,598 (62%)	1,725,000 (1%)	5,534,000 (2%)
501–10,000	11,308 (59%)	18,573 (32%)	27,322,000 (18%)	44,579,000 (19%)
More than 10,000	2,495 (13%)	3,390 (6%)	121,555,000 (81%)	192,566,000 (79%)
Total	19,236	57,561	150,602,000	242,679,000

[a]Percentage indicates the fraction of total U.S. community water supply systems in this category.
[b]Percentage is relative to the total population served by community water systems, which is less than the size of the U.S. population as a whole.

SOURCES: EPA, 1994; Public Health Service, 1965.

supply, the original authorities provided sewerage, wastewater treatment, storm water and flood control, and many other water-related services.

The comparative fragmentation of the U.S. water supply industry—and the resulting large number of water systems serving small populations—reflects the historical development pattern of the industry in the United States. Historically, each U.S. town and city developed its own local supply, subject initially to local and later to state and federal oversight. As shown in Table 1-2, ownership of small water systems was and still is fragmented among city, county, state, and federal government bodies and private utilities. The type of consolidation that occurred in England and Wales after World War II has never occurred in the United States.

INCREASING NUMBER OF WATER SUPPLY REGULATIONS

As shown in Figure 1-1, the number of drinking water contaminants regulated by the federal government has increased dramatically in the past decade, increasing the complexity for both small and large systems of providing water that meets all applicable regulations. U.S. public health officials developed the first drinking water standards in 1914, but until 1974 these standards could be legally enforced only for water transported between two or more states. The

TABLE 1-2 Ownership of U.S. Community Water Systems

	Ownership (As Percentage of Total Systems in Size Category)					
	Public					Private
Population Served	City/ Municipality	County/ Water District	State	Federal	Total Public	
Under 500	18%	7%	2%	1%	28%	72%
501–3,300	56%	17%	3%	1%	77%	23%
More than 3,300	71%	18%	2%	4%	95%	5%
Total	36%	11%	3%	1%	51%	49%

SOURCE: EPA, 1990b.

1914 standard was issued under the auspices of the Interstate Quarantine Act of 1893, which was designed to prevent the spread of communicable diseases from state to state. It initially covered only coliform bacteria, which indicate the possible contamination of water with fecal matter and thus the possible presence of disease-causing organisms (AWWA, 1990). The U.S. Public Health Service revised the standard and added new contaminants to it in 1925, 1942, 1946, and 1962. Although many states adopted the guidelines for community water supplies in their jurisdictions, the standards were federally enforceable only for municipalities whose water was used on interstate carriers such as buses, trains, airplanes, and ships (AWWA, 1990).

In 1969, a major survey of community water supplies showed that few water purveyors monitored the quality of their water to determine whether it met the recommended U.S. Public Health Service standards. Three years later, studies revealed the presence of trace quantities of contaminants in the lower Mississippi River, the source of drinking water for New Orleans (EPA, 1976). Public concern about these issues and environmental contamination in general led Congress to pass the Safe Drinking Water Act (SDWA) in 1974. The act requires that all public water supplies meet national "maximum contaminant levels" (MCLs). The EPA is responsible for developing drinking water standards under the act; the initial MCLs were based on the U.S. Public Health Service guidelines and a review of health risks of potential drinking water contaminants by the NRC (1977).

An EPA survey conducted in 1981 and 1982 of 1,000 public water systems using ground water revealed the presence of trace concentrations of volatile

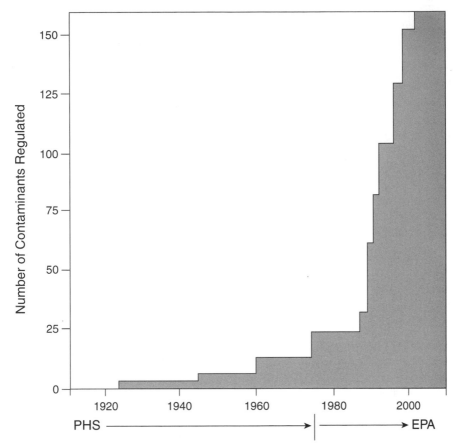

FIGURE 1-1 Number of drinking water contaminants regulated by the U.S. government. The large increase in regulated contaminants that begins after 1976 is due to regulations issued under the Safe Drinking Water Act and its amendments. SOURCE: Reprinted, with permission, from Okun (1996). ©1996 by the American Society of Civil Engineers.

organic chemicals in approximately one-quarter of the systems (AWWA, 1990). The survey results led to concern that the existing SDWA standards were insufficient to protect public health from synthetic organic chemicals in drinking water (Congressional Research Service, 1993). As a result, Congress strengthened the SDWA with a series of amendments in 1986. The amendments required the EPA to develop MCLs for 83 contaminants and, beginning in 1991, to regulate 25 additional contaminants every 3 years—thus the large increase in regulated contaminants shown in Figure 1-1. New contaminants to be regulated included a variety of volatile organic chemicals, inorganic chemicals, viruses, and parasites.

The amendments further mandated that the EPA develop rules governing which types of water supplies must provide disinfection as a minimum level of treatment and which ones must also provide filtration. Table 1-3 lists contaminants regulated by the SDWA and its amendments as of 1996. Compliance with this large number of water quality regulations can stretch the resources of the state agencies responsible for monitoring the performance of water purveyors on behalf of the federal government.

In July 1996, Congress again amended the SDWA. The amendments removed the requirement for EPA to regulate 25 new contaminants every 3 years. Instead the EPA is to regulate contaminants based on adverse health effects, occurrence, and the level of risk posed by the contaminant. In addition, the amendments established for the first time a state revolving fund to help pay for improvements to water service. Nonetheless, the states still face a major challenge in implementing the existing SDWA regulations and in determining how to respond to changes that will result from the 1996 amendments.

INCREASING RESPONSIBILITIES FOR
LOCAL AND STATE GOVERNMENTS

When Congress passed the SDWA, legislators recognized that small communities might not be able to afford the technologies needed for compliance. For example, the House of Representatives, in its summary report describing the key elements of the act, noted, "It is evident that what is a reasonable cost for a large metropolitan (or regional) public water system may not be reasonable for a small system which serves relatively few users" (Congressional Research Service, 1982). Legislators believed that the problems of small systems would self-correct. They assumed that because small communities would be unable to afford the new technologies needed to meet the regulations, small water systems would consolidate with other systems to provide a larger customer base. The House, in its legislative summary, anticipated development of "a regional water system which can afford to purchase and use [water treatment] methods, to seek additional sources of funding such as state aid, or to develop a plan for otherwise serving the affected population after any existing inadequate system is closed" (Congressional Research Service, 1982).

As indicated by the still-increasing number of small systems, this anticipated regionalization has not occurred. Instead, violations of the drinking water standards, especially the microbiological standards, have become common. For example, 29.5 percent of systems serving fewer than 500 people violated microbiological standards one or more times between October 1992 and January 1995 (see Chapter 2). Further, in implementing the SDWA, the EPA has recognized that it is not always feasible to hold small communities to the same compliance schedules, standards, and monitoring requirements as larger communities. For example, systems serving fewer than 10,000 people are not currently required to meet the MCL for trihalomethanes (THMs), created when chlorine used for disinfection reacts with naturally occurring or other organic matter in the water (EPA, 1990a).[1]   In addition, systems serving 3,300 or fewer people were given 2 extra years to complete needed treatment installations for lead contamination. For a variety of contaminants, EPA allows up to five small systems to pool their water samples and have them tested as a composite in the laboratory, rather than requiring that each system test its water individually.   Finally, exemptions from water quality standards are available to systems that cannot meet the standards due to severe economic constraints.   Such exemptions may be granted for an unlimited number of 3-year periods for systems with 500 or fewer service connections.

Although the drinking water standards are federally mandated, the states (with the exception of Wyoming[2]) are responsible for enforcing them, and inadequate enforcement capability may have contributed to the large number of violations of the drinking water standards and the lack of consolidation of water systems. Staffing levels of state drinking water programs have traditionally been constrained, and they are becoming even more stretched with the promulgation of new federal regulations.   In a 1992 resource needs survey, the Association of State Drinking Water Administrators (ASDWA) determined that, in total, state agencies had 2,272 staff years of resources available each year to implement the safe drinking water program but that they would need 4,958 staff years of effort in 1993 to implement the expanding array of federal drinking water regulations (ASDWA, personal communication, 1995).   A state program field staff person could be responsible for monitoring the operational performance and regulatory compliance of between 150 and 250 public water systems.   Assuming an equal distribution of the approximately 190,000 public water systems in the country, the average field person would be responsible for overseeing the water supply systems of more than 165 communities, both large and small.

Funding for implementation of state drinking water programs historically has been limited (and is the reason for the shortage of personnel).   A 1988 study conducted by ASDWA revealed that state drinking water programs were experiencing a $41 million budget shortfall (ASDWA and EPA, 1989).   In 1993, the U.S. General Accounting Office (GAO) concluded that "Severe resource constraints have made it increasingly difficult for many states to effectively carry out the monitoring, enforcement, and other mandatory elements of EPA's drinking water program . . . . The situation promises to deteriorate further" (GAO, 1993).

---

[1]Small systems may be required to meet THM standards in the future if the EPA implements a proposed rule governing disinfection and disinfection byproducts. Under this proposed rule, small systems would have to meet the THM standards but would not have to monitor for THMs as frequently as larger systems.

[2]All states except Wyoming have assumed primacy to enforce the SDWA (GAO, 1994).

TABLE 1-3 Contaminants Regulated Under the Safe Drinking Water Act

Individually regulated contaminants
Arsenic
Copper[a]
Fluoride
Lead[a]
Total coliforms
Total trihalomethanes (TTHM)[b]

Phase I contaminants
Benzene
Carbon tetrachloride
1,2-Dichloroethane
1,1-Dichloroethylene
p-Dichlorobenzene
1,1,1-Trichloroethane
Trichloroethylene
Vinyl chloride

Surface water treatment
Giardia lamblia[a]
Legionella[a]
Heterotrophic plate count[a]
Turbidity[a]
Viruses[a]

Phase II contaminants
Acrylamide[a]
Alachlor (Lasso)
Asbestos
Atrazine
Barium
Carbofuran
Cadmium
Chlordane
Chromium
Dibromochloropropane (DBCP)
o-Dichlorobenzene
cis-1,2-Dichloroethylene
trans-1,2-Dichloroethylene
1,2-Dichloropropane
2,4-D
2,4,5-TP (Silvex)
Ethylbenzene
Ethylene dibromide (EDB)
Epichlorohydrin[a]
Heptachlor
Heptachlor epoxide
Lindane (BHC-gamma)
Mercury
Methoxychlor
Monochlorobenzene

Phase V contaminants
Antimony
Beryllium
Cyanide
Dalapon
Di(2-ethylhexyl)adipate
Di(2-ethylhexyl)phthalate
Dichloromethane
Dinoseb
Dioxin (2,3,7,8-TCDD)
Diquat
Endothall
Endrin
Glyphosate
Hexachlorobenzene (HCB)
Hexachlorocyclopentadene
Nickel
Oxamyl (vydate)
Polycyclic aromatic hydrocarbons
   (benzo(a)pyrene)
Picloram
Simazine
Thallium
1,2,4-Trichlorobenzene
1,1,2-Trichloroethane

	Radionuclides
Nitrate	Beta particle and photon radioactivity[b]
Nitrite	Gross alpha particle activity
Polychlorinated	Radium-226
biphenyls	Radium-228
Pentachlorophenol	
Selenium	
Styrene	
Tetrachloroethylene	
Toluene	
Toxaphane	
Xylenes (total)	

NOTE: The phases in the table indicate the different stages in which the EPA issued regulations for these contaminants under the Safe Drinking Water Act and its amendments. The EPA issued regulations for phase I contaminants in 1987, for phase II contaminants in 1991, and for phase V contaminants in 1992. Regulations for surface water treatment were issued in 1989. Radionuclide regulations were issued in 1993.

[a]These contaminants are regulated by a treatment technique instead of an MCL.
[b]Regulations for these contaminants are not applicable to small systems.

SOURCE: EPA, 1994.

More recently, the EPA estimated that states would require $311 million in 1995 to implement the SDWA (Leiby, 1995). However, during 1995 the federal appropriation to states for the SDWA was $70 million, and state revenues for implementing the act totalled $126 million, leaving the states with a $115 million shortfall (Leiby, 1995; National Conference of State Legislatures, 1995). It is as yet unclear how the 1996 SDWA amendments will affect state staffing and resource requirements.

In summary, the problems of small water supply systems, while long recognized, have not been solved. The number of small systems has increased over the past three decades, despite legislation to improve water service. Ensuring that small systems are complying with the complex array of drinking water regulations is a major task for state and local governments. Addressing the small systems problem will require consideration of technical, financial, and institutional options, as discussed in this report.

REFERENCES

AWWA (American Water Works Association). 1990. Water Quality and Treatment: A Handbook of Community Water Supplies, Fourth Edition. New York: McGraw-Hill, Inc.

ASDWA (Association of State Drinking Water Administrators) and EPA (Environmental Protection Agency). 1989. State Costs of Implementing the 1986 Safe Drinking Water Act Amendments: Results and Implications of the 1988 Association of State Drinking Water Administrators Survey of State Primacy Program Resource Needs. Washington, D.C.: ASDWA.

Congressional Research Service. 1982. A Legislative History of the Safe Drinking Water Act. Serial No. 97-9. Washington, D.C.: Library of Congress, Congressional Research Service, Environment and Natural Resources Policy Division.

Congressional Research Service. 1993. A Legislative History of the Safe Drinking Water Act Amendments 1983–1992. Washington, D.C.: Library of Congress, Congressional Research Service, Environment and Natural Resources Policy Division.

EPA. 1976. Industrial Pollution of the Lower Mississippi River in Louisiana. Dallas: EPA Region VI.

EPA. 1990a. Environmental Pollution Control Alternatives: Drinking Water Treatment for Small Communities. EPA/625/5-90/025. Cincinnati: EPA, Risk Reduction Engineering Laboratory.

EPA. 1990b. Improving the Viability of Existing Small Drinking Water Systems. EPA 570/9-90-004. Washington, D.C.: EPA, Office of Water.

EPA. 1994. The National Public Water System Supervision Program: FY 1993 Compliance Report. EPA 812-R-94-001. Washington, D.C.: EPA, Office of Water.

EPA. 1995. Unpublished data from the Safe Drinking Water Information System. Washington, D.C.: EPA.

GAO (General Accounting Office). 1993. Drinking Water Program: States Face Increased Difficulties in Meeting Basic Requirements. Washington, D.C.: U.S. General Accounting Office.

GAO. 1994. Drinking Water: Stronger Efforts Essential for Small Communities to Comply with Standards. Washington, D.C.: U.S. General Accounting Office.

Leiby, V. 1995. Testimony to the Senate Appropriations Subcommittee on VA, HUD, and Independent Agencies in Regard to FY 1996 Appropriations for the PWSS Program on Behalf of the Association of State Drinking Water Administrators. Washington, D.C.: ASDWA.

National Conference of State Legislatures. 1995. Alternative Funding Mechanisms for State Drinking Water Programs, 1994–1995. Denver: National Conference of State Legislatures.

National Research Council. 1977. Drinking Water and Health. Washington, D.C.: National Academy Press.

Okun, D. A. 1977. Regionalization of Water Management: A Revolution in England and Wales. London: Applied Science Publishers.

Okun, D. A. 1996. From cholera to cancer to cryptosporidiosis. Journal of the Environmental Engineering Division, American Society of Civil Engineers 122(6):453–458.

Public Health Service. 1965. Statistical Summary of Municipal Water Facilities in the United States, January 1, 1963. Public Health Service Publication No. 1039. Washington, D.C.: Public Health Service.

Task Force on Drinking Water Construction Funding and Regionalization. 1991. Safety on Tap: A Strategy for Providing Safe, Dependable Drinking Water in the 1990s. Portland: Oregon Health Division, Drinking Water Section.

2

Status of Small Water Systems

More than 50 million U.S. residents (nearly 20 percent of the population) obtain their water from water utilities serving fewer than 10,000 people (EPA, 1994). The communities that rely on these smaller water systems are responsible for providing the financial means to build the systems, operate and maintain them, and ensure that they meet federal drinking water standards. This chapter discusses the financial status of small communities, their track record in meeting requirements of the Safe Drinking Water Act (SDWA), and their ability to pay for needed improvements to water treatment systems and infrastructures. It also reviews existing data on outbreaks of waterborne disease in small and large communities. As discussed in this chapter, small communities are often ill-equipped to assume the financial and managerial responsibilities associated with providing high-quality water service.

FINANCIAL RESOURCE LIMITATIONS IN SMALL COMMUNITIES

While every town and city is unique and not all face the same problems, many small communities have economic characteristics that make it difficult for them to raise the funds needed for adequate drinking water service. Small communities can generally be divided into two groups: those in nonmetropolitan areas and those in the outlying suburbs of major metropolitan communities.

Financial resources are typically most limited in nonmetropolitan communities. Small, nonmetropolitan communities, on average, have low per capita incomes compared to larger urban communities. Incomes averaged $38,233 in metropolitan areas in 1990 (Bureau of Census, 1990). In contrast, average incomes were $25,785 in nonmetropolitan communities with fewer than 1,000 residents, $28,872 in nonmetropolitan communities with 1,000 to 2,499 residents, and $29,192 in nonmetropolitan communities with 2,500 to 9,999 residents.

Small nonmetropolitan communities also tend to have higher unemployment rates and a larger proportion of aging residents than urban communities. The average unemployment rate in urbanized areas in 1990 was 4.9 percent, while in communities of 2,500 to 9,999 located outside of urbanized areas the average unemployment rate was 7.5 percent (Bureau of Census, 1990). A study in Virginia showed that throughout the 1980s, unemployment in nonmetropolitan areas was consistently more than 50 percent higher than in metropolitan areas (Virginia Water Project, 1994). This same study showed that 13 percent of the residents of nonmetropolitan areas in the state were over age 65, as compared to 9 percent of the population in metropolitan areas. The age demographics of small nonmetropolitan communities reflect a historical trend of younger generations increasingly migrating to larger cities in search of jobs. For example, in 1880, 75 percent of Americans lived in rural areas, but now 75 percent live in urban areas (Lindsey, 1995).

Adding to the financial difficulties of small nonmetropolitan communities, lenders are less willing to loan to rural communities than to metropolitan ones because of the increased effort needed to monitor smaller loans relative to the profits they generate. Rural banks often prefer to invest in government securities rather than in local efforts because of the need to diversify their risks (Lindsey, 1995). A shortage of loan capital is an especially significant problem for privately owned small water systems because they are not eligible to receive the government grants available to some publicly owned systems.

Some small water systems are located in or near metropolitan areas, where they could be, but often are not, connected to a major municipal water supply. An example of such a system is one that until recently served the town of Aroma Park, Illinois, which is 50 miles from Chicago and only 2 miles from a major water utility serving cities south of Chicago (see Box 2-1). Small water systems located in metropolitan communities are becoming an increasingly common phenomenon as city residents migrate to new housing developments in suburban and periurban areas. The private developers who build these new communities may purposefully avoid acquiring water service from the central city in order to save on development costs. The water purveyor is initially the developer, but once the development is complete, responsibility shifts to a homeowners' association, which may be poorly equipped to manage water service. An example of a region with a large number of small systems in a metropolitan area is Kitsap County, Washington (see Box 2-2).

Whether a small community is located in a metropolitan area or a nonmetropolitan one, it will lack the economies of scale of larger communities in providing water service. The small ratepayer base available in small communities means

BOX 2-1 Small Water Supply System Near Chicago

The Village of Aroma Park, Illinois, is located approximately 50 miles south of Chicago. Prior to 1944, the source of water for this municipal system, which serves approximately 700 people, was two wells. During periods of low precipitation, this supply was diminished, and the village was forced to institute water use restrictions. The village water system investigated additional ground water supplies, but these other ground water sources were not adequate. The cost to provide a surface water plant was excessive.

In 1994, the village and Consumers Illinois Water Company (CIWC), an investor-owned water utility using a major surface water source and serving the nearby larger communities of Kankakee, Bradley, and Bourbonnias (combined population of approximately 60,000), worked together to extend a 12-in.-diameter main approximately 2 miles from the main CIWC system to the Aroma Park system. The village now receives a reliable, ample supply of water from CIWC and has retained its employees and the ownership of its system. The project, which cost approximately $300,000, was funded half by a state grant obtained by the village and half by CIWC. In addition to residents of Aroma Park, a number of rural and suburban residents along the route of the new main have connected to the new main and enjoy improved water service and fire protection.

that per capita water rates must be higher than in larger, urban communities in order to provide the same level of service. Small pipes, tanks, and pumps cost more per unit of water delivered than the larger sizes. For example, the per capita capital cost of a conventional water treatment plant with a maximum flow capacity of 0.01 m^3/s (0.23 million gallons per day), which would be adequate for a population of about 1,500, is more than three times as high as the per capita cost for a system with a maximum flow capacity of 0.1 m^3/s (2.3 million gallons per day), which would be adequate for a population of approximately 15,000 (McMahon, 1984; Montgomery, 1985). Thus, residents of very small communities might pay more than three times as much to finance construction of a new water treatment plant as residents of larger towns. Similarly, paying staff to operate and maintain small water systems is much more costly per unit of water delivered than paying staff to run larger systems. In one small Connecticut community, capital and operating improvements to a failing water system would have increased residential water rates nearly sevenfold per year, from $144 to $1,000. When this system merged with a large one, the community's residents obtained better quality water for $269 per year instead of $1,000 per year because their capital and operation and maintenance costs were spread across a much larger service area (EPA, 1989a).

The limitations in financial resources available to many small communities for water service create a variety of problems. In the most extreme instances, residents of small communities may lack running water altogether. As of 1990,

---

BOX 2-2 Proliferation of Urban Small Systems in Kitsap County, Washington

Washington State has experienced almost a 24 percent increase in population over the last 10 years. This rapid growth has occurred primarily in proximity to urban centers, but largely in areas where services and infrastructures were not available. While Washington has been a leader in promoting comprehensive planning and the concept of water service areas with sole providers, there has been a proliferation of small systems over the past ten years, not only in the rural areas but even more so in the suburban areas and urban fringes.

Kitsap County is one example of a highly populated area where the number of small water supply systems has increased in recent years. The county is one of four that make up the greater Puget Sound metropolitan area, which includes the urban corridor from Everett (Snohomish County) to Tacoma (Pierce County), with Seattle (King County) in between. Although land-use patterns indicate that a large part of Kitsap County is rural and forested, the county is second only to King County in population density of all counties in Washington State; in 1995, the population density was 582 persons per square mile.

Despite Kipsap County's proximity to major metropolitan centers and its high population density, the number of regulated water systems that must meet SDWA requirements (15 connections or more) increased from 216 in 1987 to 255 in 1995. Water systems with 3 to 14 connections increased from 453 in 1987 to 810 in 1995.

Factors driving the increase in small water systems in Kitsap County include the costs of connecting to an existing system and, perhaps more importantly, state water allocation policies. Water rights are required for larger systems to increase their allocations to provide additional service connections. They are not required for smaller systems. It currently can take up to 5 years for allocation decisions (yes or no) to be made. This state water resource policy failure has had severe impacts on attempts to stop the proliferation of small systems, particularly in areas that have above-average county population densities.

In 1995, the state legislature passed a bill that requires new water systems to be owned or managed by satellite operators certified by the state if the community cannot be served by an adjacent, preexisting utility. This should help stop the proliferation of small systems.

---

more than 1.1 million U.S. households lacked plumbing, according to the U.S. Census Bureau. Of these 1.1 million households, approximately 760,000 are located in communities with populations of less than 10,000. Lack of water service may result not just from the absence of household plumbing and water mains to the community but also from seasonal inadequacies in the water source and infrastructure failure because of poor operation and maintenance. Low water pressure, caused by an inadequate or intermittent supply, may have serious consequences for water quality because it can allow contaminants to infiltrate the water through leaks in the pipes. Backsiphonage of contaminants into water pipes during periods of low pressure has caused dozens of waterborne disease

---

**BOX 2-3 Water Supply Problems in a
Small Pennsylvania Town**

Residents of the Village of Onnalinda, in Cambria County, Pennsylvania, have had to boil their drinking water for more than a year. The village water system provides water to residents from a reservoir on a tributary of the Little Conemaugh River. Until the chlorinator broke, the water was disinfected just before it entered the village. However, when the chlorinator broke and the village could not afford to replace it, residents were advised to boil their water. The water system ceased all water quality monitoring at this time, claiming lack of financial resources. The utility claimed it would also be unable to afford installation of a filtration system, which is now required for all water systems using surface water sources. In addition to these problems, many of the pipes in the water system are old and require replacement.

A 1994 study by the Redevelopment Authority of Cambria County concluded that Onnalinda's water supply problems could best be solved by connecting to a nearby larger water utility, the Highland Sewer and Water Authority (Pellegrini Engineers, 1994). The study estimated the total costs of this project at $286,000, or nearly $16,000 for each of the village's 18 homes. The Highland Sewer and Water Authority concluded that in order to take over the system, it would need a grant or other attractive offer because on its own the Onnalinda system would not generate enough income to pay for the project.

---

outbreaks over the past two decades (Craun, 1996). While data on water quality failures are readily available, data on service failures are not routinely acquired by any agency.

While most small communities in the United States do have access to running water, they often cannot afford to construct the facilities or maintain the staff needed to ensure compliance with the SDWA. In addition, small communities may be unable to replace corroded distribution piping and other failing or substandard infrastructure. Further, they may lack the resources needed to develop and carry out the detailed planning necessary for long-term improvements to their water service. The town described in Box 2-3 illustrates these types of problems.

COMPLIANCE WITH THE SAFE DRINKING WATER ACT IN SMALL COMMUNITIES

The limited resources available to small communities can create difficulties in complying with the SDWA. The SDWA regulates two broad types of contaminants: microbiological and chemical. Analysis of U.S. Environmental Protection Agency (EPA) data on SDWA compliance shows that small communities, especially those with fewer than 500 people, often have difficulty in meeting the requirements for microbiological contaminants and have more difficulty than large communities in meeting requirements for chemical contaminants.

Compliance with Microbiological Standards

Waterborne diseases—caused by microorganisms that enter water sources from the wastes of infected humans and animals—have been largely but not entirely eliminated in the United States and other industrialized countries, thanks in part to better protection of water sources and wider use of water treatment systems. While in recent years the public and federal policymakers have placed more emphasis on chemical contaminants than on microbiological ones, publicity surrounding a 1993 epidemic in Milwaukee brought national attention to the risk of waterborne pathogens. An estimated 403,000 Milwaukee residents contracted cryptosporidiosis (severe, prolonged diarrhea caused by the parasite *Cryptosporidium*) via the city's drinking water supply (MacKenzie et al., 1994). Public health investigators estimated that more than half of those who obtained their water from the contaminated supply became ill (MacKenzie et al., 1994), and deaths due to cryptosporidiosis were reported among patients with compromised immune systems. Health investigators estimated that this outbreak cost the city more than $133 million in direct medical costs (such as those for hospital and clinical treatments) and indirect medical costs (such as those associated with lost wages) (P. A. Shaffer, Centers for Disease Control, personal communication, 1996).

The microbiological quality of drinking water is regulated under the SDWA by requiring water systems to monitor for coliform bacteria, which indicate the possible presence of fecal contamination and disease-causing organisms (see Box 2-4). Table 2-1 shows the number of community water systems that violated the maximum contaminant level (MCL) for total coliforms by size of community and water source (ground or surface) for the 27-month period October 1, 1992, through December 31, 1994.[1] Most of the systems in violation were in ground water systems serving 500 or fewer people, presumably because many of these systems do not disinfect their water. The violation rate for systems with fewer than 500 customers is more than twice the rate for systems serving larger populations: a violation of the MCL for total coliforms was reported by 29.5 percent of the systems serving fewer than 500 people as compared to less than 14.5 percent of the systems serving larger communities.

The large number of small community water systems in violation of the SDWA poses a serious management problem for state regulatory agencies. As

---

[1]A large number of systems fail to follow EPA requirements for contaminant monitoring, so the actual number of systems in violation could be much higher than shown in Table 2-1. In 1993, for example, 26 percent of water systems serving 500 or fewer people, 20 percent of those serving 501 to 3,300 people, 37 percent of those serving 3,301 to 10,000 people, and 26 percent of those serving larger communities violated one or more of EPA's monitoring and reporting requirements (EPA, 1994).

> **BOX 2-4 Monitoring the Microbiological Quality of Drinking Water**
>
> A wide variety of bacteria, viruses, and parasites can cause illness when present in drinking water. However, because monitoring for each of the possible waterborne disease agents is technically difficult, coliform bacteria have been used for many decades as an indicator of the microbiological quality of drinking water. Rather than requiring water systems to monitor for all possible types of waterborne pathogens, SDWA regulations require that systems check for the presence of coliform bacteria. These bacteria are present in the normal intestinal flora of humans and other warm-blooded animals and are found in large numbers in fecal wastes. Most species of coliforms are also free-living in the environment. Thus, their presence in drinking water does not necessarily represent fecal contamination. However, finding coliform bacteria in a drinking water system indicates possible fecal contamination due to inadequate water treatment or deficiencies in the distribution system.
>
> In 1977, the EPA issued regulations for total coliforms in drinking water, establishing an MCL based on coliform density, monitoring requirements, and analytical method used (EPA, 1976). A revised regulation, which became effective on December 31, 1990 (EPA, 1989b), specifies an MCL based on the presence of coliforms in a 100-ml water sample: coliform bacteria can be detected in no more than 5.0 percent of the samples collected during a month; systems collecting fewer than 40 samples per month may have no more than one positive sample without violating the MCL. The revised regulation specifies a monitoring frequency based on the number of people served and requires additional monitoring whenever a positive sample occurs. For example, systems serving 25 to 1,000 people need collect only one coliform sample per month, while systems of 1,001 to 2,500 and 2,501 to 3,300 consumers must collect two and three samples per month, respectively. A maximum of 480 samples per month is required of systems serving more than 3,960,000 people.

shown in Table 2-1, 96 percent (13,039 of 13,526) of the systems in violation serve communities with 10,000 or fewer people.

While it is generally presumed that coliform-free water contains few or no pathogens and is therefore unlikely to cause waterborne disease, the risk of infectious waterborne disease may be greater than is suggested by the number of violations of the MCL for total coliforms. Waterborne disease outbreaks, especially those caused by disinfectant-resistant organisms such as *Giardia* and *Cryptosporidium*, have occurred in water systems that have not violated the coliform MCL (Craun, 1984, 1990a; Moore et al., 1994; Kramer et al., 1995). For example, the 1993 outbreak in Milwaukee and an outbreak in Las Vegas in 1994 occurred even though both cities were in compliance with the coliform MCL (MacKenzie et al., 1994; Goldstein et al., 1996). Thus, federal data on violations of coliform standards may understate the degree to which drinking water supplies are contaminated with pathogenic organisms.

TABLE 2-1 Number of Community Water Systems that Violated the
Maximum Contaminant Level for Total Coliform Bacteria Between October 1,
1992, and December 31, 1994

	Size of Population Served by Water System				
	500 and Under	501– 3,300	3,301– 10,000	10,001 and Over	Total
Surface water systems	524	294	167	199	1,184
Ground water systems	9,985	1,644	425	288	12,342
Total number of systems with violations	10,509	1,938	592	487	13,526
Percentage of systems with violations	29.5	13.4	14.4	14.4	23.5

SOURCE: Federal Reporting Data System (data summaries provided by Jeff Sexton, EPA).

Compliance with Chemical Standards

A major factor leading to passage of the SDWA was public concern about contamination of the environment with man-made chemicals that, over long periods of exposure, can lead to cancer and other chronic diseases. Thus, the SDWA requires the EPA to set MCLs for dozens of chemical contaminants. With the exception of nitrate and nitrite, which can cause acute methemoglobinemia, the MCLs for chemicals are based on the prevention of adverse health effects associated with long-term, low-level exposures (see Box 2-5).

Tables 2-2, 2-3, and 2-4 show the number of community water systems with violations of chemical MCLs for the 27-month period October 1, 1992, through December 31, 1994. Table 2-2 shows violations by individual contaminant for systems using ground water as a water source; Table 2-3 shows violations for systems using surface water; and Table 2-4 summarizes the data from these tables and shows the percentage of systems in violation for various sizes of water systems.

As documented in Table 2-4, the likelihood that a water system will violate chemical MCLs is a function of the size of the system. The violation rate for systems serving 500 or fewer people is more than triple the rate for systems serving populations of greater than 10,000 people. While the rate of violation of chemical standards appears to be low based on these data, the actual violation rates are likely higher than the data show because violations are not always reported. According to EPA data, 11 percent of systems serving 500 or fewer

BOX 2-5 Assessing Risks of Chemical Contaminants

Understanding the nature of risks associated with long-term, low-level exposures to chemical contaminants requires knowledge of how MCLs are established. Although human health effects data are used when available, this information is sparse for most chemicals. The MCLs are derived almost exclusively from toxicological studies on animals in combination with an analysis of the technical feasibility of monitoring for and removing the contaminant in question.

Risk assessments used to establish health effects of a chemical are based on one of two approaches, depending on whether the chemical is considered to cause cancer (i.e., every dose has a risk) or not (i.e., there is a threshold dose below which no effect is expected). For chemicals not suspected to cause cancer, either the dose at which no detectable adverse health effect occurs or the lowest dose at which an adverse effect is detected can be used in establishing an MCL. For carcinogens, MCLs are generally based on the estimated risk of one additional cancer per 100,000 to 1 million people exposed over a lifetime.

A number of uncertainties are associated with these risk assessments, including the quality and quantity of the available health data, extrapolation of toxicological data from high to low doses and from animals to humans, and assumptions used to derive the quantitative risk estimates (ingestion of 2 liters of water per day is assumed for a lifetime of 70 years, and exposure from drinking water must be estimated relative to exposure from other sources). Thus, the MCLs, although they are legal requirements, should be considered only crude indications of the actual health risk associated with long-term exposure to any chemical. It should be recognized that the use of large uncertainty factors for nonthreshold chemicals and conservative estimates for cancer risks generally ensures that moderate short-term exposure to levels exceeding the MCL will not significantly increase the risk of disease.

customers failed to meet monitoring and reporting requirements for chemical contaminants in 1994–1995; approximately 6 percent of systems serving 501 to 10,000 customers failed to meet these requirements (EPA, 1995).

Table 2-2 shows that for small systems using ground water, the most common individual chemical contaminant is nitrate. Of the 646 small ground water systems with reported violations, 463 (72 percent) had problems meeting the standards for nitrate. The second leading problem for these systems was fluoride, which was reported as a problem by 57 (8.8 percent) of the small ground water systems with violations. As a group, man-made organic chemicals (including all of the contaminants included in phase I of the SDWA regulations and some of those included in phase II) were reported as problems by 81 (13 percent) of the small ground water systems with violations.

Table 2-3 shows that for small systems using surface water, the leading chemical contamination problem is atrazine. Atrazine violations were reported by 37 (51 percent) of the 72 small surface water systems with violations. Nitrate was the second leading cause of chemical contamination for small surface water

TABLE 2-2 Number of Ground Water Systems that Violated Drinking Water Standards for Chemical Contaminants Between October 1, 1992, and December 31, 1994

	Size of Population Served by Water System			
Chemical	500 and Under	501–3,300	3,301–10,000	10,001 and Over
Individually regulated contaminants				
Arsenic	5			
Fluoride	48	9		1
Lead (prior to distribution system)	1			
Phase I contaminants				
Benzene	8	1		
Carbon tetrachloride	3			
1,2-Dichloroethane	2			
1,1-Dichloroethylene	7	3		
1,1,1-Trichloroethane	4			
Trichloroethylene	14	3	3	1
Vinyl chloride	3	1		
Phase II contaminants				
Barium	6	5		
Cadmium	2	1		
1,2-Dichloropropane	3		1	
Ethylene dibromide	3		1	
Monochlorobenzene			1	
Nitrate	369	87	7	2
Nitrite	1			
Selenium	20	6		
Tetrachloroethylene	13	3	2	1
Total number of systems with violations	512	119	15	5

NOTE: Only contaminants for which violations occurred are included on this list. Also, contaminants for which regulations did not yet apply to small systems during the period covered by this table are excluded from the list.

SOURCE: Federal Reporting Data System (data summaries provided by Jeff Sexton, EPA).

TABLE 2-3 Number of Surface Water Systems that Violated Drinking Water
Standards for Chemical Contaminants Between October 1, 1992, and
December 31, 1994

	Size of Population Served by Water System			
Chemical	500 and Under	501–3,300	3,301–10,000	10,001 and Over
Individually regulated contaminants				
Fluoride	1	2		
Phase I contaminants (VOC rule)				
1,1,1-Trichloroethane		1		
Trichloroethylene	1	2		
Phase II contaminants				
Atrazine	8	23	6	1
Chromium				1
Ethylene dibromide	2	1		
Mercury				1
Nitrate	7	13	4	6
Tetrachloroethylene		1		1
Total number of systems with violations	19	43	10	10

NOTE: Only contaminants for which violations occurred are included in this list. Also, contaminants for which regulations did not yet apply to small systems during the period covered by this table are excluded from the list.

SOURCE: Federal Reporting Data System (data summaries provided by Jeff Sexton, EPA).

TABLE 2-4 Violations of Drinking Water Standards for Chemical
Contaminants by Size of Water System: Summary Data for the Period October
1, 1992, to December 31, 1994

	Size of Population Served by Water System			
Source of Water	500 and Under	501–3,300	3,301–10,000	10,001 and Over
Ground water	512	119	15	5
Surface water	19	43	10	10
Total number of systems with violations	531	162	25	15
Percentage of systems with violations	1.5	1.1	0.61	0.44

SOURCE: Tables 2-2 and 2-3.

systems, reported by 24 (33 percent) of the small surface water systems with violations. Thus, while small systems, on average, violate the chemical standards for drinking water more frequently than larger systems, from the monitoring data available the range of contaminants affecting these systems appears fairly narrow.

HEALTH RISKS OF INADEQUATE
DRINKING WATER TREATMENT

What are the health risks of failing to comply with SDWA regulations? While no precise number can be assigned to this risk due to limitations of existing data, an analysis of the data is instructive because it reveals that outbreaks of waterborne disease are not uncommon.

More than 600 outbreaks of waterborne disease have been reported to the federal government since 1971, but the actual incidence of waterborne disease is likely much higher than this. Data on waterborne disease outbreaks originate from state and local public health agencies, which conduct disease surveillances and investigate outbreaks, some of which they determine to be waterborne. Reporting of waterborne disease outbreaks by these state and local agencies to the federal government is voluntary. The EPA and Centers for Disease Control (CDC) maintain a national surveillance program to track waterborne disease outbreaks that are reported to the federal government. Since 1971, the CDC and EPA have periodically reported waterborne outbreak statistics (CDC, 1973, 1974, 1976a,b, 1977, 1979, 1980, 1981, 1982a,b, 1983, 1984, 1985; St. Louis, 1988; Levine et al., 1990; Herwaldt et al., 1991; Moore et al., 1993; Kramer et al., 1996a,b). These reports form the basis for current assessments of waterborne disease in the United States. The EPA and CDC use information from epidemiologic investigations to determine the causes of these outbreaks (the contaminants involved and whether the outbreak originated with the water source or the distribution system).

Figure 2-1 shows the national distribution of waterborne disease outbreaks reported to the federal government between 1976 and 1992. The figure shows outbreaks by size of the community where the outbreak occurred and by whether the outbreak resulted from a distribution system problem or contamination of the water source. The period 1976 to 1992 is used because of the availability of census data to determine the community size. As shown in the figure, waterborne disease outbreaks have been reported in communities of all sizes in every region of the country.

Table 2-5 shows microbiological disease agents responsible for most of the waterborne disease outbreaks for the period 1971 through 1994, according to CDC data. With the exception of hepatitis A, all of these cause diarrhea. Many are also associated with nausea and fever. Hepatitis A causes chronic liver problems that may result in symptoms such as fever, tiredness, loss of appetite,

FIGURE 2-1 Waterborne disease outbreaks reported and documented to EPA and CDC in the United States between 1976 and 1992. The states reporting the most outbreaks are not necessarily the ones in which the most outbreaks occurred. Recognition and investigation of possible outbreaks depend on factors such as physician interest, consumer awareness, extent of local health department surveillance activities, and aggressiveness of health

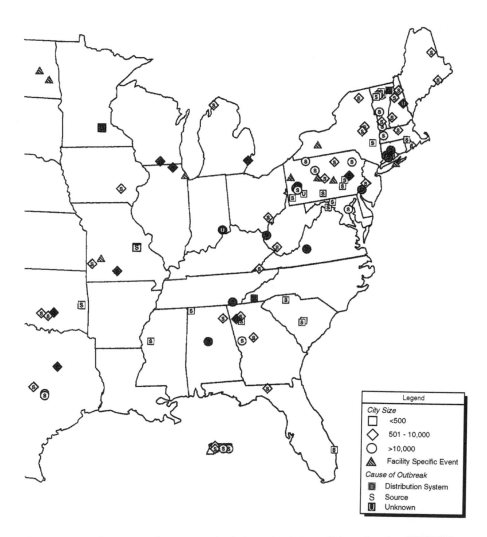

departments and water regulatory agencies in investigating possible outbreaks. SOURCE: Complied from CDC and EPA data in CDC, 1977, 1979, 1980, 1981, 1982a,b, 1983, 1984, 1985; St. Louis, 1988; Levine et al., 1990; Herwaldt et al., 1991; Moore et al., 1993.

TABLE 2-5   Infectious Waterborne Disease Outbreaks Caused by
Contaminated Drinking Water Systems in the United States, 1971–1994

Etiologic Agent	Type of Organism	Outbreaks	Cases of Illness
Giardia lamblia	Parasite	113	26,670
Shigella	Bacteria	40	8,962
Norwalk-like virus	Virus	20	10,552
Hepatitis A	Virus	28	827
Campylobacter	Bacteria	15	5,459
Salmonella	Bacteria	13	2,995
Cryptosporidium parvum	Parasite	10	419,934[a]
All others (toxigenic *E. coli* bacteria, *Yersinia* bacteria, rotavirus, *S. typhi* bacteria, *V. cholerae* bacteria, and others)		15	3,263
Unidentified etiologies		325	81,554
TOTAL		579	560,421

[a]Of these cases, 403,000 were from a single outbreak in Milwaukee.

SOURCE: Compiled from CDC and EPA data in CDC, 1973, 1974, 1976a,b, 1977, 1979, 1980, 1981, 1982a,b, 1983, 1984, 1985; St. Louis, 1988; Levine et al., 1990; Herwaldt et al., 1991; Moore et al., 1993; Kramer et al., 1996a,b.

and nausea followed by jaundice. As shown in the table, the pathogen causing the illness is unidentified for more than half of the outbreaks.

Table 2-6 shows incidents of acute chemical poisoning caused by drinking water contamination between 1971 and 1994. Copper and fluoride were the most frequently identified chemicals causing acute illness—usually vomiting within a short time after consuming the water or using it in beverages. The source of copper in these outbreaks was leaching from plumbing and water service pipes due to corrosive water (a problem that can be controlled with proper water treatment). The source of fluoride was various operational and technological deficiencies in the equipment used to fluoridate the water to prevent dental problems. Two deaths occurred as a result of high fluoride levels. Other deaths included two attributed to arsenic contamination of a private well, one caused by high levels of nitrate in a private well, and one resulting from high levels of ethylene

TABLE 2-6 Chemicals Associated with Waterborne Outbreaks of Acute Illness in the United States, 1971–1994

Chemical	Community Water Systems	Noncommunity Water Systems	Individual Water Systems	Total Number of Outbreaks
Copper	13	2	2	17
Fluoride	7	5	0	12[a]
Nitrate	0	0	7	7
Oils	4	1	0	5
Chlordane	4	0	0	4
Lead	1	0	5	6
Ethylene glycol	2	0	0	2
Sodium hydroxide	2	0	0	2
Arsenic	0	1	1	2
Unidentified herbicide	2	0	0	2
All others[b]	7	0	5	12
TOTAL	42	9	20	71

[a]Includes an outbreak where high levels of both fluoride and copper were present.

[b]Single outbreaks caused by ingestion of selenium, phenol, benzene, PCB, gasoline, chromium, trichlorethylene, ethyl acrylate, morpholine, or hydroquinone and a dermatitis outbreak due to high levels of chlorine.

SOURCE: Compiled from CDC and EPA data in CDC, 1973, 1974, 1976a,b, 1977, 1979, 1980, 1981, 1982a,b, 1983, 1984, 1985; St. Louis, 1988; Levine et al., 1990; Herwaldt et al., 1991; Moore et al., 1993; Kramer et al., 1996a,b.

TABLE 2-7 Causes of Waterborne Outbreaks of Acute Chemical Illness in the United States, 1971–1994

Type of Water System	Cause of Outbreak			
	Ground Water Contamination	Treatment Deficiencies	Distribution or Storage Deficiencies	Unidentified Deficiencies
Community	4	7	29	2
Noncommunity	1	4	3	1
Individual	12	1	5	2
All systems	17	12	37	5

SOURCE: Compiled from CDC and EPA data in CDC, 1973, 1974, 1976a,b, 1977, 1979, 1980, 1981, 1982a,b, 1983, 1984, 1985; St. Louis, 1988; Levine et al., 1990; Herwaldt et al., 1991; Moore et al., 1993; Kramer et al., 1996a,b.

glycol from a cross-connection. As shown in Table 2-7, 52 percent of acute illnesses due to chemical contamination were caused by contamination of the water distribution system, either by cross-connections, backsiphonage, or corrosive water. In addition, 24 percent of the outbreaks were caused by ground water contamination, often with nitrate.

While these data show that waterborne disease is still a problem in the United States, health experts generally agree that the number of reported outbreaks represents only the tip of the iceberg. For a variety of reasons, the scope and quality of the available data are highly variable. Problems with the interpretation of these data include the following:

• Some states have weak disease surveillance systems. As a consequence, waterborne disease outbreaks may have occurred in these states without being recognized and investigated by state officials.

• Even where adequate disease surveillance systems are in place, health officials may not recognize small outbreaks or recognize that water is the route of disease transmission for endemic diseases. When low levels of waterborne pathogens are present, sporadic cases of infection may occur at various times and points throughout the water distribution system and never be recognized as an outbreak because it is difficult to separate these cases from background levels of disease. Only a small proportion of people who develop diarrheal illness, which is sometimes caused by waterborne infections, seek physician assistance.

• Not all outbreaks are adequately investigated. Outbreaks are included in the EPA and CDC data base only if water quality and/or epidemiologic data are collected to document water as the route of disease transmission.

• States do not always report waterborne disease outbreaks to the CDC. As mentioned above, reporting of disease outbreaks to the federal government is voluntary.

• The data do not reflect the risks of long-term, low levels of exposure to chemical contamination, only the cases of acute chemical poisoning. Consumption of water with levels of a carcinogen higher than the established MCL can theoretically increase the potential cancer risk for the exposed population, especially as the duration of exposure and concentration of the carcinogen increase. However, scientists are usually unable to accurately quantify this risk for the specific population that may be exposed. The same is true for potential teratogens.

Although public health officials acknowledge that the incidence of waterborne disease is much higher than the reported waterborne outbreak statistics indicate, the actual incidence is difficult to estimate. Few studies have attempted to quantify the extent of underreporting of waterborne outbreaks and associated disease. Limited evidence suggests that anywhere from 50 to 90 percent of waterborne outbreaks may be unreported (Craun and McCabe, 1973; Craun,

1990b). Furthermore, the extent to which endemic disease is associated with drinking water is unknown. An epidemiologic study in the Montreal area attributed 35 percent of mild, unreported cases of diarrhea to the consumption of tap water that met water quality regulations (Payment et al., 1991). In Vermont, waterborne transmission was suggested to be an important cause of non-outbreak-related giardiasis, as rates of infection were almost twice as high in persons receiving unfiltered versus filtered municipal water (Birkhead and Vogt, 1989).

Although the number of waterborne disease outbreaks is greater for small systems, available data do not indicate whether small water systems experience a higher percentage of outbreaks than larger systems (Craun and McCabe, 1973; G. F. Craun, Global Consulting for Environmental Health, unpublished data, 1993). Analyses of differences in the occurrence of waterborne disease between small and large communities are complicated by outbreak and disease reporting differences. Detection and investigation of outbreaks may differ between large and small communities. In small communities, for example, waterborne disease outbreaks may not be recognized because of the small number of people affected.

ADEQUACY OF WATER TREATMENT AND DISTRIBUTION SYSTEMS IN SMALL COMMUNITIES

Public health officials have long advocated a "multiple barrier" approach to disease prevention, and drinking water treatment and distribution systems are key components of this approach. The multiple barriers to waterborne disease include

- selection of the purest sources of water;
- protection of both ground and surface water sources from municipal, agricultural, and industrial pollution;
- appropriate treatment of drinking water;
- effective operation and monitoring of drinking water treatment facilities; and
- prevention of contamination during the storage and distribution of treated water.

As the U.S. population continues to increase and put pressure on natural resources, finding high-quality source water will become more difficult, and water treatment systems will increase in importance as a barrier to waterborne illnesses. Failures and weaknesses in this barrier leave small communities more vulnerable to outbreaks of waterborne disease.

Many small water systems require major upgrades to their treatment and distribution systems. A recent EPA survey of 600 small systems selected at random from around the United States revealed major deficiencies in drinking water treatment (Fraser et al., 1995). Some of the surveyed systems using surface water disinfect but do not filter their water, putting consumers at significant risk of exposure to waterborne pathogens, particularly disinfectant-resistant protozoa such as *Giardia* and *Cryptosporidium*. Many of the small surface water systems that do provide filtration use turbid sources of water with inadequate filtration. For example, one system in Wyoming must issue boil water advisories for long periods each spring because its filtration system is not adequate to purify the extremely turbid spring runoff water. Ground water systems using water with high concentrations of iron and manganese often lack adequate equipment for removal of these metals, which can cause foul tastes and odors and clog pipes. In some small ground water systems, high concentrations of total dissolved solids make the water unpalatable or even undrinkable.

The EPA survey also found that maintenance of water treatment and distribution systems is often poor in small communities. The survey found frequent instances of facilities that, although within their service life, are inoperable because of lack of maintenance. In many of the small systems surveyed, equipment such as chemical pumps, turbidimeters, and pH monitors is broken. Water distribution lines are often substandard or have not been replaced for many decades or even a century. One system distributes water through garden hoses. Others have small-diameter galvanized steel water mains that are completely corroded. Maintenance in some plants is so poor that the entire plant needs to be replaced to make the system reliable.

Improving deteriorating or inadequate water treatment and distribution facilities is the responsibility of local authorities, whether village, town, city, or county governments, or, in the case of private developments, the owner of the development. Many small communities lack the resources to initiate, let alone carry out, the steps necessary to upgrade their water service and would be obliged to call on the state for assistance even when only a simple treatment technology is required. However, as discussed in Chapter 1, most state agencies lack the resources to provide the detailed technical assistance necessary at the local level. As a consequence, small communities with poorly performing water systems can have difficulty in obtaining the help they need to improve their water service. In addition, state regulators may be unable to provide the oversight needed to ensure that small water systems are properly maintained.

Diversity of Local Water Supply Agencies

Originally, most local water supply service was provided privately by entrepreneurs who saw financial opportunity in filling the need for a necessary service. However, in the late nineteenth and early twentieth centuries, as many private enterprises failed for lack of adequate capital resources to keep up with population growth, or for other reasons, local governments began to take over these utilities, especially in larger communities. As shown in Table 1-2, most very small systems are still privately owned, but most systems serving 500 or more people are now publicly owned.

Various arrangements are commonly found for providing water supply services in smaller communities:

• The water supply system is owned and operated by the local authority, which may be most commonly the incorporated community, village, town, city, county, or regional government that serves the small community systems and individual householders within its jurisdiction.

• The water purveyor is initially the developer of a suburban community, but when the development is complete, responsibility for the water service may become that of a homeowners' association of some type. The movement of populations to suburban and periurban areas has led to private developers being responsible for the creation of free-standing residential communities; developers often purposefully avoid acquiring water service from the central city for financial reasons.

• The water supply is owned and operated by the owner of a mobile home park. In such situations, the capital costs of water supply (and sewerage) facilities are a much larger fraction of the overall costs of the residences than in conventional single-family housing communities. The residents of mobile home parks, in addition, are more severely constrained in borrowing for such facilities.

• The community, which may have built and initially operated its water supply system, contracts to a private company some or all of its responsibilities for managing the system. The contracting out of specific tasks such as meter reading, billing, and analytical work to meet water quality monitoring obligations are examples of this practice.

• The water utility, including all of its capital facilities, is owned and operated by a private investor-owned company, generally one that owns and operates small systems throughout a region or throughout the United States.

• A community in the vicinity of a larger city arranges for service from that city. The larger city may provide all of the infrastructure and deal with individual customers in the outlying community from the outset. More commonly, a small community will provide its own distribution system, often following the guidelines and standards of the larger city, and arrange to purchase raw or treated water wholesale from the larger city but be responsible for dealing with individual customers. In time, the smaller community may be absorbed into the service area of the larger one.

Complexity of Improving Water Service

Regardless of who owns the water system, improving or providing new water supply service requires a series of complex steps. These steps may be beyond the capability of small communities, especially those without links to larger utilities.

The first step in improving water supply service is planning (see Chapter 5). This involves developing quantitative information about the future population, water users, water demand, current and future yield of water sources, quality of the sources, treatment methods for the sources, environmental consequences associated with the water supply (including creating the need for additional sewerage), financing, and capital and operating costs. It also requires establishment of the mechanisms for obtaining the necessary engineering services and selecting engineers for design and construction.

The second step in improving water service is designing the system improvements. Even if package plants are being considered, the community will need to retain a design engineer for tasks such as selecting the water source, designing the transmission and distribution system lines, designing the reservoir, and selecting the most appropriate package plant from amongst the many available options.

The third step in upgrading water service is constructing the new system. It is common for a community consulting engineer to help negotiate the contract for construction of the facilities with a general contractor. In cases where all design and construction work is purchased in a package, a community representative or paid staff person must supervise construction to ensure that the design and specifications are fully met by the contractor.

The final step in upgrading water service is to provide for adequate operation, maintenance, and management of the system. Large communities usually employ large staffs—including a manager, one or more engineers, a chemist and/or microbiologist, laboratory technicians, one or more plant operators, maintenance personnel for the distribution system and treatment facilities, meter readers, clerical personnel, customer service representatives, and drivers—to manage their water systems. While the chemistry and microbiology involved in small water supply systems are little different from those of larger systems, the staff resources for small communities are far more limited. In small systems, it is not uncommon for one part-time individual to manage the system alone.

In sum, improving water supply service is a complex task that many small communities are unable to handle on their own.

CONCLUSIONS

While waterborne disease in the United States is far less prevalent now than it was at the turn of the century, U.S. public water supplies still face significant difficulties. Small communities face special problems in supplying high-quality drinking water because many of them cannot afford the technologies or trained personnel needed to meet federal drinking water standards. As the population continues to increase and sources of clean water become harder to find, the safeguards provided by properly operated, modern water treatment technologies will become increasingly important in communities of all sizes. Yet, small communities are often ill equipped to arrange for improvements to their water service. A sustained national effort is needed to ensure that small communities have the support they need to provide an adequate quantity of water that meets the standards of the SDWA. In particular, policymakers need to address the following issues:

• **Small communities have economic characteristics that make it difficult for them to raise the funds needed for adequate water supply service.** Small communities, especially those in nonmetropolitan areas, may have relatively low per-capita incomes, high unemployment rates, large populations of elderly residents, and limited access to capital for loans. In addition, because of their small ratepayer bases, per-person water rates in small communities must be higher than those in large communities in order to provide the same level of service.

• **Small water systems, especially the very small ones, have difficulty complying with SDWA requirements for microbiological and chemical contaminants.** Systems serving fewer than 500 people exceeded MCLs more than twice as often as those serving more than 10,000 people.

• **Many small water systems require major capital investments to upgrade their treatment systems and infrastructure.** Many small systems not only lack the treatment facilities needed to meet regulatory requirements but also have broken equipment or corroded or substandard distribution lines that need to be replaced. These small systems could benefit from technical assistance from state water supply regulators, but state agencies generally lack the resources to provide the detailed assistance that would most benefit small systems.

• **Failure to provide adequate water treatment and to comply with drinking water standards leaves small communities vulnerable to outbreaks of waterborne illnesses.** More than 600 outbreaks of waterborne disease have been reported in the United States in the past two decades; these reported outbreaks represent only a fraction of the total incidence of waterborne disease because many outbreaks are unrecognized or unreported.

REFERENCES

Birkhead, G., and R. S. Vogt. 1989. Epidemiologic surveillance for endemic *Giardia lamblia* infection in Vermont. American Journal of Epidemiology 129(4):762-768.

Bureau of Census. 1990. 1990 Census of Population: Social and Economic Characteristics. Washington, D.C.: Department of Commerce.

CDC (Centers for Disease Control). 1973. Foodborne Outbreaks, Annual Summary, 1972. Publication No. (CDC) 74-8185. Washington, D.C.: Department of Health, Education, and Welfare.

CDC. 1974. Foodborne and Waterborne Outbreaks, Summary, 1973. Publication No. (CDC) 75-8185. Washington, D.C.: Department of Health, Education, and Welfare.

CDC. 1976a. Foodborne and Waterborne Outbreaks, Summary, 1974. Publication No. 76-8185. Washington, D.C.: Department of Health, Education, and Welfare.

CDC. 1976b. Foodborne and Waterborne Outbreaks, Summary, 1975. Publication No. 76-8185. Washington, D.C.: Department of Health, Education, and Welfare.

CDC. 1977. Foodborne and Waterborne Outbreaks, Summary, 1976. Publication No. 78-8185. Washington, D.C.: Department of Health, Education, and Welfare.

CDC. 1979. Foodborne and Waterborne Disease, 1977. Publication No. 79-8185. Washington, D.C.: Department of Health, Education, and Welfare.

CDC. 1980. Water-related Disease Outbreaks, Summary, 1978. Publication No. 80-8385. Washington, D.C.: Department of Health and Human Services.

CDC. 1981. Water-related Disease Outbreaks, Summary, 1979. Publication No. 81-8385. Washington, D.C.: Department of Health and Human Services.

CDC. 1982a. Water-related Disease Outbreaks, Summary, 1980. Publication No. 82-8385. Washington, D.C.: Department of Health and Human Services.

CDC. 1982b. Water-related Disease Outbreaks, Summary, 1981. Publication No. 82-8385. Washington, D.C.: Department of Health and Human Services.

CDC. 1983. Water-related Disease Outbreaks, Summary, 1982. Publication No. 83-8385. Washington, D.C.: Department of Health and Human Services.

CDC. 1984. Water-related Disease Outbreaks, Summary, 1983. Publication No. 84-8385. Washington, D.C.: Department of Health and Human Services.

CDC. 1985. Water-related Disease Outbreaks, Summary, 1984. Publication No. 99-2510. Washington, D.C.: Department of Health and Human Services.

Craun, G. F. 1984. Waterborne outbreaks of giardiasis: current status. In Giardia and Giardiasis, S. L. Erlandsen and E. A. Meyer, eds. London: Plenum Press.

Craun, G. F. 1990a. Waterborne giardiasis. In Giardiasis, E. A. Meyer, ed. Amsterdam: Elsevier Science Publishers.

Craun, G. F. 1990b. Review of the causes of waterborne outbreaks. Pp. 1–22 in Methods for the Investigation and Prevention of Waterborne Disease Outbreaks. EPA/600/1-90/005a. Washington, D.C.: Environmental Protection Agency.

Craun, G. F. 1996. Waterborne disease in the United States. Pp. 55–77 in Water Quality in Latin America, G. F. Craun, ed. Washington, D.C.: ILSI Press.

Craun, G. F., and L. J. McCabe. 1973. Review of the causes of waterborne-disease outbreaks. Journal of the American Water Works Association 65(1):74.

EPA (Environmental Protection Agency). 1976. National Interim Primary Drinking Water Regulations. EPA 590/9-76-003. Washington, D.C.: EPA.

EPA. 1989a. Ensuring the Viability of New, Small Drinking Water Systems: A Study of State Programs. EPA-57019-89-004. Washington, D.C.: EPA.

EPA. 1989b. Drinking Water: National Primary Drinking Water Regulations, Total Coliforms. 40 CFR Parts 141, 54 (124):27–544 and 142, 54 (124):27–568.

EPA. 1994. The National Public Water System Supervision Program FY 1993 Compliance Report. EPA 812-R-94-001. Washington, D.C.: EPA, Office of Water.

EPA. 1995. Unpublished data from the Safe Drinking Water Information System. Washington, D.C.: EPA. September 20.

Fraser, D., C. Davies, and R. T. Jones. 1995. Capital needs of small systems. Journal of the American Water Works Association 87(11):32–38.

Goldstein, S. T., D. D. Juranek, O. Ravenholt, A. W. Hightower, D. G. Martin, J. L. Mesnik, S. D. Griffiths, A. J. Bryant, R. R. Reich, and B. L. Herwaldt. 1996. Cryptosporidiosis: an outbreak associated with drinking water despite state-of-the art water treatment. Annals of Internal Medicine 124(5):459–468.

Herwaldt, B. L., G. F. Craun, S. L. Stokes, and D. D. Juranek. 1991. Waterborne-disease outbreaks, 1989–1990. Morbidity and Mortality Weekly Report 40(SS-3):1–22.

Kramer, M. H., B. L. Herwaldt, G. F. Craun, R. L. Calderon, and D. D. Juranek. 1996a. Surveillance for waterborne-disease outbreaks–U.S., 1993–1994. Morbidity and Mortality Weekly Report 45(SS-1):1–33.

Kramer, M. H., B. L. Herwaldt, G. F. Craun, R. L. Calderon, and D. D. Juranek. 1996b. Waterborne disease: 1993 and 1994. Journal of the American Water Works Association 88(3):66–80.

Levine, W. C., W. I. Stephenson, and G. F. Craun. 1990. Waterborne disease outbreaks, 1986–1988. Morbidity and Mortality Weekly Report 39(SS-1):1–13.

Lindsey, L. B. 1995. The Future in Rural America. Presentation at the Renaissance of Rural America Conference, Memphis, Tenn., March 7.

MacKenzie, W. R., N. J. Hoxie, M. E. Proctor, M. S. Gradus, K. A. Blair, D. E. Peterson, J. J. Kazmierczak, D. G. Addiss, K. R. Fox, J. B. Rose, and J.P. Davis. 1994. A massive outbreak in Milwaukee of Cryptosporidium infection transmitted through the public water supply. New England Journal of Medicine 331:161–167.

McMahon, L. A. 1984. 1984 Dodge Guide to Public Works and Heavy Construction, Annual Edition No. 16. Princeton, N.J.: McGraw-Hill Information Systems Company.

Montgomery, J. M., Consulting Engineers. 1985. Water Treatment Principles and Design. New York: John Wiley & Sons.

Moore, A. C., B. L. Herwaldt, G. F. Craun, R. L. Calderon, A. K. Highsmith, and D. D. Juranek. 1993. Surveillance for waterborne disease outbreaks–U.S., 1991–1992. Morbidity and Mortality Weekly Report 42(SS-5):1–22.

Moore, A. C., B. L. Herwaldt, G. F. Craun, R. L. Calderon, A. K. Highsmith, and D. D. Juranek. 1994. Waterborne-disease outbreaks in the United States, 1991 and 1992. Journal of the American Water Works Association 84(2):87–99.

Payment, P., L. Richardson, J. Siemiatycki, R. Dewar, M. Edwards, and E. Frances. 1991. A randomized trial to evaluate the risk of gastrointestinal disease due to consumption of drinking water meeting current microbiological standards. American Journal of Public Health 81(6):703–8.

Pellegrini Engineers. 1994. Small Water Systems Feasibility Study in Connection with the Highland Sewer and Water Authority for Redevelopment Authority of Cambria County. Altoona, Pa.: Pellegrini Engineers.

St. Louis, M. E. 1988. Water-related disease outbreaks, summary, 1985. Morbidity and Mortality Weekly Report 37(SS-2):15-24.

Virginia Water Project. 1994. Rural Virginia: A Profile in Diversity. Roanoke, Va.: Virginia Water Project.

3

Technologies for the Small System

Most source waters used for public drinking water supplies are not of suitable quality for consumption without some form of treatment. The U.S. Environmental Protection Agency (EPA) has ruled that all surface waters must be filtered and disinfected before consumption unless the purveyor can justify avoidance of filtration; some surface waters also need to be treated with additional processes to remove chemical contaminants before they are suitable for use as drinking water. Many ground water sources are disinfected, and many are treated to remove nuisance chemicals (such as iron and manganese) and chemical contaminants before distribution. This chapter evaluates water treatment processes that can be used by small systems and discusses their suitability under various conditions.

The fundamental responsibility of a public water system is to provide safe drinking water, as defined by the Safe Drinking Water Act (SDWA) and its amendments. Water utilities are required by the SDWA to monitor drinking water quality. When source water used by a water system does not meet quality requirements, the utility has several options. The first that should be considered is finding a cleaner, safer source water that requires less treatment than the existing source water, for this is often the most cost- and resource-efficient way to meet demand. Surface water sources tend to be turbid and typically contain higher concentrations of colloidal and microbiological material than ground water sources. Ground water sources generally have higher initial quality and tend to require less treatment than surface water sources, making ground water sources a good choice for small water systems. In fact, as shown in Table 3-1, most small systems already use ground water sources. Before installing new treatment systems, a small utility using surface water might seek a ground water source, or a

TABLE 3-1 Water Source for Community Water
Systems of Various Sizes

	Water Source	
Population Served	Ground Water	Surface Water
Small systems		
Under 500	91%	9%
501–3,300	74%	26%
3,301–10,000	58%	42%
Large systems		
10,001–100,000	46%	54%
More than 100,000	28%	72%

SOURCE: EPA, 1994.

utility using a poor ground water source might develop a new well in an alternative location or use a deeper aquifer by extending the depth of a well or drilling a deeper one. In either case, if alternative sources of high-quality raw water are not available, the utility might seek a source of treated water from a water utility that has an adequate supply of water and is located close enough to extend a transmission main at an affordable cost. If such options cannot be found, however, then the utility needs to explore adding additional treatment systems.

TREATMENT TECHNOLOGIES: OVERVIEW

Table 3-2 lists treatment processes according to the water quality problems they address. No single process can solve every water quality problem. Rather, a utility must choose from a wide range of processes that are used for different purposes. The treatment technology or combination of technologies to be used in a specific situation depends on the source water quality, the nature of the contaminant to be removed, the desired qualities of the treated water, and the size of the water system. For very small systems, treatment may not be a feasible alternative because of the high cost of having a treatment system designed and installed and the complexity of maintaining it.

Historically, the design of drinking water treatment systems has been driven by the need to remove microbial contaminants and turbidity. Microbial contaminants are the central concern because they can lead to immediate health problems. Turbidity is a concern not only because water containing particles can have an objectionable taste and appearance but also because particles of fecal matter can harbor microorganisms, and soil particles can carry sorbed contaminants such as pesticides and herbicides. Aesthetic problems such as excess hardness, which

TABLE 3-2 Treatment Technologies by Contaminant Type

	Disinfectants/Oxidants						Air Stripping Systems	
	Free Cl$_2$	NH$_2$Cl	ClO$_2$	O$_3$	Ultraviolet Radiation	KMnO$_4$	Aeration	Membrane Aeration
General water quality parameters								
Turbidity								
Color						X		
Disinfection byproduct precursors								
Taste and odor				X	X	X	X	X
Biological contaminants								
Algae				X				
Protozoa			X	X				
Bacteria	X	X	X	X	X			
Viruses	X	X	X	X	X			
Organic chemicals								
Volatile organic compounds (VOCs)							X	X
Semivolatile compounds								
Pesticides							X	X
Biodegradable organic matter								

Inorganic chemicals					
Hardness (calcium and magnesium)					
Iron	X	X	X	X	X
Manganese	X	X	X	X	X
Arsenic					
Selenium					
Thallium					
Fluoride					
Radon				X	X
Radium					
Uranium					
Cations					
Anions					
Total dissolved solids					
Nitrate					
Ammonia					

continued on next page

TABLE 3-2 *Continued*

| | Adsorption Systems | | | |
	Powdered Activated Carbon	Granular Activated Carbon	Ion Exchange	Activated Alumina
General water quality parameters				
Turbidity	X	X		
Color	X	X		
Disinfection byproduct precursors	X	X		
Taste and odor	X	X		
Biological contaminants				
Algae		X		
Protozoa		X		
Bacteria		X		
Viruses		X		
Organic chemicals				
VOCs	X	X		
Semivolatile compounds	X	X		
Pesticides	X	X		
Biodegradable organic matter	X	X		

Inorganic chemicals			
Hardness		X	
Iron			
Manganese			
Arsenic			X
Selenium			X
Thallium			X
Fluoride		X	X
Radon			
Radium		X	
Uranium			
Cations		X	
Anions		X	
Total dissolved solids	X		
Nitrate		X	
Ammonia		X	

continued on next page

TABLE 3-2 *Continued*

	Membrane Processes				
	Microfiltration	Ultrafiltration	Nanofiltration	Reverse Osmosis	Electrodialysis/ Electrodialysis Reversal
General water quality parameters					
Turbidity	X				
Color		X	X	X	
Disinfection byproduct precursors		X	X	X	
Taste and odor		X	X		
Biological contaminants					
Algae	X	X	X	X	
Protozoa	X	X	X	X	
Bacteria		X	X	X	
Viruses			X	X	
Organic chemicals					
VOCs					
Semivolatile compounds				X	
Pesticides			X	X	
Biodegradable organic matter					

continued on next page

Inorganic chemicals

Hardness	X	X	X
Iron			X
Manganese			X
Arsenic		X	X
Selenium		X	X
Thallium		X	X
Fluoride			X
Radon			
Radium		X	X
Uranium		X	X
Cations		X	X
Anions		X	X
Total dissolved solids		X	X
Nitrate			X
Ammonia			

56

TABLE 3-2 Continued

	Filtration Systems						
	Direct Filtration	Conventional Filtration	Dissolved Air Flotation	Diatomaceous Earth Filtration	Slow Sand Filtration	Bag/ Cartridge Filters	Lime Softening
General water quality parameters							
Turbidity	X	X	X	X	X		X
Color	X	X	X				
Disinfection byproduct precursors	X	X	X				
Taste and odor					X		
Biological contaminants							
Algae	X	X	X	X	X		
Protozoa	X	X	X	X	X	X	X
Bacteria	X	X	X	X	X		X
Viruses	X	X	X	X	X		X
Organic chemicals							
VOCs							
Semivolatile compounds							
Pesticides							
Biodegradable organic matter	X[a]	X[a]		X[a]	X[a]		

Inorganic chemicals

Hardness	X				X
Iron	X	X	X	X	X
Manganese	X	X	X	X	X
Arsenic	X	X			X
Selenium	X				X
Thallium					
Fluoride					
Radon					
Radium	X				X
Uranium					
Cations	X				X
Anions					
Total dissolved solids					
Nitrate					
Ammonia	X[a]	X[a]	X[a]		X[a]

[a]Operated in biologically active mode.

can lead to scaling of water heaters and excess soap consumption, and objection-able tastes and odors have also played an important historical role in the develop-ment of drinking water treatment technologies. Finally, the corrosivity of the water has been a longstanding concern because of the need to protect water mains and plumbing. Drinking water treatment systems are still designed primarily with these objectives in mind rather than being based on the need to remove trace levels of synthetic chemicals to comply with requirements of the SDWA and its amendments.

Because so many regulations apply to drinking water, small systems must look at the entire spectrum of drinking water regulations before deciding on a treatment method. The system manager who considers the regulations and other water quality concerns on a piecemeal basis can end up using first one process and then another until finally the treatment plant becomes a costly chain of processes inefficiently tacked on to one another. Eventually the small system could find that it can no longer afford to install further treatment systems, and the whole investment might be made for naught.

A number of the treatment processes listed in Table 3-2 and described in more detail below are available to small communities as package plants. The term "package plant" is not intended to convey the concept of a complete water treatment plant in a package. Rather, a package plant is a grouping of treatment processes, such as chemical feed, rapid mixing, flocculation, sedimentation, and filtration, in a compact, preassembled unit. To provide a complete treatment plant, other equipment, or in some cases a series of package plants, generally is required. For example, most package plants designed to provide water filtration are not also equipped with equipment for disinfection, corrosion control, or ad-sorption of organic contaminants by granular activated carbon (GAC).

Some manufacturers prefer to call package plants "preengineered" process equipment because the process engineering for the package plant design has been done by the manufacturer. What remains for the water system's engineer to design is the specifics of the on-site application of the equipment. Because package plants do not require custom design, and because the process facilities (for example, mixing chamber, flocculation basin, sedimentation basin, and fil-ter) are built in a factory instead of on site, such systems have the potential to provide significant cost savings to small communities.

Table 3-3 outlines important capital considerations for common water treat-ment processes. Water treatment technologies change constantly. As shown in the table, at any given time they fall into one of several broad categories:

• *Conventional technologies* are in widespread use and familiar to practic-ing treatment engineers and operators.
• *Accepted technologies* are not as widely used as conventional technolo-gies. Sometimes these technologies have been developed for other fields and adopted by the water community. Some processes of this type have performed

satisfactorily in water treatment, but some personnel in the field may not be familiar with them.

• *Emerging technologies* include those that have not been applied to water treatment in an operating system but show great promise for acceptance in the near future. These technologies are likely to be in the research or pilot plant stage.

Table 3-3 also shows the costs of different treatment processes on a relative scale. Precise cost information cannot be provided because costs change constantly. For example, advances in membrane processes are reducing the costs of membrane systems.

Table 3-4 shows operating considerations—raw water quality, operator skills, monitoring requirements, and costs—for common treatment processes. As shown in the table, different treatment processes have different requirements for source water quality. Some processes require "high-quality" or "very-high-quality" source water. Details about quality requirements are provided with the individual technology descriptions later in this chapter.

Once a treatment system has been selected and installed, it is common to believe that the major expenditure is over. This is true for relatively few technologies. Operation and maintenance costs must be considered in long-term planning and in selection of treatment processes because they vary with the technology, as shown in Table 3-4. Skill levels required of water treatment plant operators also vary with system complexity and type of technology. Table 3-4 indicates different skill levels:

• In a *basic* system, an operator with minimal experience in the water treatment field can perform the necessary system operation and monitoring if provided with proper instruction. The operator is capable of reading and following explicit directions but would not necessarily have water treatment as a primary career.

• In an *intermediate* system, the operator needs to understand the principles of water treatment and have a knowledge of the regulatory framework. The operator must be capable of making system changes in response to source water fluctuations.

• In an *advanced* system, the operator must possess a thorough understanding of the principles of system operation. The operator should be knowledgeable in water treatment and regulatory requirements, with water treatment being the career objective. (The operator may, however, have advanced knowledge of only the particular treatment technology.) This operator seeks information, remains informed, and reliably interprets and responds to water fluctuations and system intricacies.

Tables 3-3, 3-4, and others in this chapter are meant only to guide prelimi-

TABLE 3-3 Capital Considerations for Treatment Technologies

Technology	Contaminants	State of Technology	Relative Capital Cost
All water sources			
Disinfection	Microbiological contaminants		
Free Cl_2		Conventional	Low
NH_2Cl		Conventional	Low
ClO_2		Accepted	Low
O_3		Accepted	Medium
Ultraviolet radiation		Accepted	Medium
Corrosion control	Prevention of system corrosion, lead, copper		
Chemical feeders		Conventional	Low
Limestone contactor		Accepted	Medium
Membrane filtration systems	Turbidity, protozoa (*Giardia* and *Cryptosporidium*), algae, and the following:		
Microfiltration	Some bacteria	Accepted	Medium
Ultrafiltration	Bacteria, some viruses, some color	Accepted	Medium
Nanofiltration	Bacteria, viruses, color, some organic chemicals, hardness	Emerging	Medium
Reverse osmosis	Bacteria, viruses, humic acids, some organic chemicals, inorganic chemicals, hardness, radium, salts	Conventional	Medium
Electrodialysis/electrodialysis reversal	Inorganic chemicals (charged)	Accepted	High

Adsorption			
Powdered activated carbon (PAC)	Organic chemicals, tastes and odors	Conventional	Low/medium
Granular activated carbon (GAC)	Organic chemicals, tastes and odors, microorganisms	Accepted	Medium/high
Lime softening	Hardness, iron, manganese, turbidity	Accepted	High
Ground water sources			
Aeration			
Diffused air	Volatile organic chemicals, radon, tastes and odors	Accepted	Low
Mechanical aeration	Volatile organic chemicals, radon, tastes and odors	Accepted	Low
Tray aerators	Volatile organic chemicals, radon, tastes and odors	Conventional	Low/medium
Packed tower aeration	Volatile organic chemicals, radon, tastes and odors	Conventional	Medium
Membrane aeration	Volatile and semivolatile organic chemicals radon, tastes and odors	Emerging	Medium
Oxidation/filtration			
Permanganate	Reduced iron and manganese, organic chemicals, radon, tastes and odors	Conventional	Low
O_3	Reduced iron and manganese, organic chemicals, tastes and odors	Accepted	High

continued on next page

TABLE 3-3 *Continued*

Technology	Contaminants	State of Technology	Relative Capital Cost
Ion exchange	Inorganic chemicals, radium, nitrate	Accepted	Medium
Activated alumina	Arsenic, thallium, selenium, fluoride, other inorganic chemicals	Accepted	High
Surface water sources			
Coagulation-filtration	Turbidity, color, disinfection byproduct precursors, microorganisms, algae, iron, manganese, biodegradable organic matter,[a] ammonia[a]		
Direct filtration		Accepted	High
Conventional, with sedimentation		Conventional	High
Dissolved air flotation		Accepted	High
Diatomaceous earth filtration	Turbidity, algae, *Giardia, Cryptosporidium*, biodegradable organic matter,[a] ammonia[a]	Accepted	Medium/high
Slow sand filtration	Turbidity, microorganisms, biodegradable organic matter,[a] ammonia,[a] tastes and odors		
Uncovered filters		Conventional in some states	Medium
Covered filters		Conventional in some states	Medium/high
Bag filters and cartridge filters	*Giardia* cysts and *Cryptosporidium* oocysts	Accepted in some states	Low

[a]If operated in biologically active mode.

nary consideration of a treatment process. Reference to the textual description of the process later in this chapter is also necessary to further assess its applicability to a given water system.

In the descriptions that follow, treatment processes are grouped according to whether they are suitable for small systems using either surface water or ground water, are best suited to ground water systems, or are used primarily for surface water systems.

TECHNOLOGIES FOR ALL SYSTEMS

Contamination with microorganisms is common to surface water sources and is becoming an increasing concern for ground water sources. Other water quality concerns common to both surface and ground water systems are excess corrosivity, hardness, and, increasingly, contamination with synthetic organic chemicals. The technologies described in this section address these water quality concerns, as well as some others, and are suitable for use in treating either surface water or ground water.

Disinfection

How the Process Works

Disinfection is the inactivation of pathogens in drinking water. Although not entirely effective against all pathogens, disinfection is the most cost-effective way to reduce the incidence of waterborne disease. Two common techniques are chemical disinfection and irradiation with ultraviolet (UV) light.

UV disinfection is used primarily in small systems that treat ground water. UV radiation has been demonstrated to be effective against bacteria and viruses, which are the microbiological contaminants likely to be found in ground waters for which the quality is not directly influenced by surface water. However, because it does a poor job of killing *Giardia* and *Cryptosporidium*, UV radiation is not an accepted means for disinfecting surface waters, unless they have already been treated in a way that would physically remove the cysts and oocysts of the *Giardia* and *Cryptosporidium*.

The chemical disinfectants used in drinking water treatment are free chlorine, chloramine, ozone, and chlorine dioxide. Iodine has been studied as a disinfectant, but the EPA restricts its use to short-term, limited, or emergency purposes because of concerns over possible adverse health effects such as iodine hypersensitivity and thyroid problems (EPA, 1982, 1995).

Of the chemical disinfectants, free chlorine is probably used most commonly, with chloramine next in popularity. In a survey conducted in 1989 and 1990, approximately 72 percent of the nearly 280 water utilities responding reported using free chlorine (AWWA Committee, 1992). Approximately 21 per-

TABLE 3-4 Operational Considerations for Treatment Technologies

Technology	Raw Water Quality Range[a]
All water sources	
Disinfection	All, but better with higher quality
Free Cl_2	
NH_2Cl	
ClO_2	
O_3	
Ultraviolet radiation	
Corrosion control	
Chemical feeders	All ranges
Limestone contactor	Low iron, low turbidity
Membrane filtration systems	
Microfiltration	Needs high water quality (or pretreatment)
Ultrafiltration	Needs very high water quality (or pretreatment)
Nanofiltration	Needs very high water quality (or pretreatment)
Reverse osmosis	Requires prefiltration for surface water
Electrodialysis/electrodialysis reversal	Requires prefiltration for surface water
Adsorption	
Powdered activated carbon (PAC)	All waters
Granular activated carbon (GAC)	Surface water may require prefiltration
Lime softening	All waters
Ground water sources	
Air stripping	
Diffused air	All ground waters
Mechanical aeration	All ground waters
Tray aeration	All ground waters
Packed tower aeration	All ground waters
Membrane aeration	All ground waters
Oxidation/filtration	
Tray aerators	All ground waters
Permanganate	All ground waters
O_3	All ground waters
Cl_2	All ground waters
Ion exchange	All ground waters
Activated alumina	All ground waters
Surface water sources	
Coagulation-filtration	
Direct filtration	Needs high raw water quality
Conventional, with sedimentation	Can treat wide range of water quality
Dissolved air flotation	Very high algae OK, high color OK, moderate turbidity
Diatomaceous earth filtration	Needs very high water quality
Slow sand filtration	Needs very high water quality
Bag and cartridge filters	Need very high quality water

[a]Refer to text for detailed description of water quality needs.

Operator Skill Level Required	Monitoring Requirements	Relative Operating Cost
Basic	Low	Low
Basic	Low	Low
Intermediate	High	Low
Intermediate	Low	Medium
Basic	Low	Low
Basic	Low	Medium
Basic	Low	Low
Basic	Low	Low
Basic	Low	Medium
Basic	Low	Medium/high
Advanced	Medium	High
Advanced	Medium	High
Intermediate	Low	Medium/high
Basic	Low/medium	Medium/high
Advanced	High	High
Basic	Low	Low
Basic	Low	Low
Basic	Low	Low
Intermediate	Low	Medium
Intermediate	Low	Medium
Basic	Low	Low
Intermediate	Medium	Medium
Intermediate	Low	Low
Basic	Low	Low
Intermediate	Medium	Medium/high
Advanced	Medium	High
Advanced	High	Medium/high
Advanced	High	Medium/high
Advanced	High	Medium/high
Intermediate	Medium	Medium/high
Basic	Low	Low, with good raw water
Basic	Low	Low/high; depends on cycle length

cent of the utilities reported using chloramine; approximately 5 percent used chlorine dioxide in conjunction with free chlorine or chloramine. Ozone was used by approximately 0.4 percent of the utilities.

Since 1990, a considerable number of ozone facilities have come on line, although the percentage of utilities using ozone remains small compared to the percentages using chloramine or free chlorine. If ozone is used for disinfection of surface waters, the ozone can break down complex organic molecules into smaller organic molecules or molecular fragments that are more readily used by bacteria as a food source. Using ozone can thus increase the biological instability of the water and result in a higher level of bacterial growth in the distribution system. One remedy for biological instability is to employ biological filtration. This is done by using conventional filter media or GAC as a filter media in conjunction with a delay in the application of chlorine, chloramine, or chlorine dioxide until after the water is filtered. The growth of bacteria in the biological filter does not impair filtration efficacy, and some organic matter can be removed, improving the biological stability of the water. Any filtration plant that does not apply a disinfectant other than ozone before filtration is, in effect, practicing a form of biological filtration, so this practice would not be beyond the capability of small system operators.

EPA regulations require that a disinfectant residual be maintained in distribution systems of water utilities that treat surface water. UV radiation leaves no residual, and ozone dissipates too rapidly to leave a residual. Therefore, maintaining a distribution system residual requires using free chlorine, chloramine, or chlorine dioxide.

An emerging approach to disinfection involves the electrolytic generation of mixed disinfectants, which produces ozone, chlorine dioxide, and chlorine. Electrolytic equipment has been used in water treatment for at least 20 years in the United States to generate chlorine from a sodium chloride solution, but use of such equipment to generate a mixture of disinfectants is a new concept. However, such processes currently are not a suitable option for small systems to use in meeting the EPA's disinfection requirements because of the difficulty of measuring the concentrations of multiple disinfectants in water and lack of data for evaluating the effectiveness of mixed disinfectants. Measurement of ozone, chlorine dioxide, chlorite, chlorate, and free chlorine in a single sample probably is not possible outside of a chemistry research laboratory, if it can be done there. Furthermore, the effectiveness of disinfectants in inactivating bacteria, viruses, and protozoa is estimated based on empirical data, and insufficient data are available for disinfectant mixtures. In the absence of data on the performance of mixed disinfectants under a wide range of water quality conditions, this type of technology cannot be applied to meet the EPA's requirements for disinfection.

Monitoring Requirements

EPA regulations require systems to periodically monitor the residual concentration of disinfectant before water is served to the first customer on the distribution system. This regulatory requirement reflects the practical reality that monitoring is essential to successful disinfection because it provides evidence that a disinfectant residual has been attained. Without monitoring capability, an operator has no basis for knowing that disinfection is adequate. Test kits or spectrophotometers allow for easy monitoring of the disinfectant residual for free chlorine, total chlorine (free and combined), and ozone. Monitoring residual levels of chlorine dioxide and its degradation products, chlorate and chlorite, is more challenging and probably beyond the capability of most small systems.

The intensity or rigor of chemical disinfection provided in the treatment plant, before water is delivered to customers, is assessed in terms of CT, in which C is the residual concentration of the disinfectant in milligrams per liter and T is the time in minutes for which the water and disinfectant chemical were in contact. The product of these parameters is a measure of the effectiveness of disinfection and is used to determine compliance with drinking water standards.

The second factor in the CT value, contact time, depends on, among other things, the geometry of the vessel or basin containing the water to which disinfectant has been added and the rate of flow of the water through the contact basin. The EPA requires that the contact time be based on the time (T_{10}) in which the first 10 percent of water entering the basin would leave, rather than on the theoretical contact time. This is a conservative approach, but it ensures that only 10 percent of the water in the contact basin has a contact time equal to or less than the time used for assessing the value of CT. Therefore, to be able to report CT values the plant operator must also know the rate of flow at the plant and the value of T_{10}.

Because temperature influences the efficiency of disinfection, water temperature must be monitored. The values of CT required for effective disinfection increase as water temperature decreases, reflecting the experimental observation that the resistance of microorganisms to disinfectants increases by factors of 2 to 3 for each 10°C decline in temperature.

When free chlorine is used as a disinfectant, its efficacy decreases as pH increases. Therefore, monitoring the pH of the water during disinfection is important for free chlorine. The EPA's CT values for free chlorine reflect this dependence on pH.

UV disinfection devices need built-in monitors to indicate the intensity of the UV radiation. Ideally, an automatic shut-off should prevent water flow if the UV intensity is not adequate to provide the level of disinfection required.

Disinfection byproducts, harmful compounds that form when water is disinfected, will become another aspect of water quality that small water systems must monitor and manage when the EPA's proposed Disinfection/Disinfectant

Byproducts (D/DBP) Rule takes effect within a few years. While water systems serving fewer than 10,000 persons were not included in the rule that established a drinking water standard for trihalomethanes (THMs), which are common disinfection byproducts, small systems will be included in the new D/DBP Rule. The new rule will set lower limits for THMs and new standards for haloacetic acids (HAAs). Therefore, in the next century small water systems will need to use disinfection methods that are effective for killing pathogens without forming excessive disinfection byproducts. Disinfection byproduct compliance is more likely to be a problem for small water systems treating surface waters than for those treating ground waters because surface water sources tend to contain more natural organic matter that forms byproducts when mixed with disinfectants.

Formation of byproducts depends on the quality of the source water and on the disinfectant used. Free chlorine forms THMs, HAAs, and other compounds classified as disinfection byproducts. Adding ammonia to chlorinated water forms chloramine and stops formation of most byproducts. Chloramine can cause formation of cyanogen chloride, but this compound is not regulated, nor does EPA plan to regulate it in the near future. Ozone does not form chlorinated byproducts, but in some waters that contain bromide it can form bromate and brominated byproducts that will be regulated in the D/DBP Rule. Ozone also forms aldehydes, but these are not currently scheduled for regulation. Chlorine dioxide minimizes the formation of byproducts, but this disinfectant breaks down over time and forms chlorite and chlorate. The EPA plans to regulate chlorate in the future. UV radiation produces no disinfection byproducts that are of concern at the present time.

When disinfection byproducts are regulated for small water systems, systems that use a disinfectant other than UV radiation will need to monitor for these products in their distribution systems. Small systems planning to begin use of disinfection will need to evaluate byproduct formation to be sure that they will not create regulatory compliance problems from the disinfection techniques they are planning to use.

Operating Requirements

Of all operating requirements, the most critical aspect for any disinfection process is that it MUST operate whenever drinking water is produced. This is especially true for disinfection systems used in conjunction with filtration processes, such as bag filters and cartridge filters, that are not capable of removing viruses and most bacteria. Any disinfection system intended to function in the absence of a plant operator should include automatic monitoring devices that shut down the plant if disinfection becomes inadequate. Such cases require that an adequate treated water supply be on hand when the water system is shut down, or that a "boil water" order be issued.

Routine tasks for a plant operator include monitoring disinfectant residual, maintaining disinfectant feed equipment, and ensuring that an adequate supply of disinfectant is on hand when chlorine or chloramine are used. When chlorine dioxide is used, adequate quantities of feedstocks must be kept. Operators of systems using ozone need to maintain the ozone generation and air preparation equipment.

When chloramine is used, both chlorine and ammonia must be added to the water. This can be done with solution feeders for calcium hypochlorite or sodium hypochlorite and ammonium sulfate. Liquid chlorine in cylinders that provide chlorine gas under pressure can also be used, although its use is not favored by some systems because of transportation and storage hazards. With chloramine, chemicals must be fed accurately. If the ratio of chlorine to ammonia falls outside of the appropriate range, water quality problems can arise in the distribution system, either from production of dichloramine and nitrogen trichloride (which can cause odor problems) if chlorine is overfed or from the presence of ammonia (which can lead to biological instability of the water) if ammonia is overfed. While chloramine use provides important advantages in the distribution system, particularly with respect to minimizing disinfection byproduct formation, chloramination must be monitored and controlled carefully. In addition, chloramine is not as strong a disinfectant as chlorine, so it requires a much higher CT value.

Generation of chlorine dioxide is more complex than production of chloramine. Because of this complexity, as well as the complexity of monitoring, chlorine dioxide may not be appropriate for most small systems.

A shortcoming of many small systems, particularly those with package plants, is the small amount of disinfection contact time (T) available. To reduce capital costs, many small systems do not have the extensive storage needed to ensure the proper contact time, particularly when water temperatures are near freezing. Opferman et al. (1995), in a paper that assessed CT compliance in Ohio, reported, "In Ohio, several small operators elected to close their treatment plants and link with a larger countywide water supply system rather than invest in clearwell upgrades." Meeting the CT requirement may be a major challenge for some small systems, particularly those that use chloramine.

A second shortcoming for many water systems, both small and large, is that chemical disinfectants (mainly chlorine and chloramine, the most widely used disinfectants) are often added to water without provision for thorough and rapid mixing into the water being treated. Much greater care is used to mix coagulant chemicals in water than to mix chlorine into water, yet both can accomplish their intended functions only after they have been dispersed into all of the water. The past practice of adding chlorine to water without much forethought as to how it was mixed may reflect lax attitudes toward disinfection in the era before *Giardia* (i.e., to the end of the 1970s), when maintaining a free chlorine residual of 0.2 mg/liter at the end of 30 minutes (a CT of 6) was considered an adequate disinfection practice. This practice would no longer be acceptable for surface water treatment.

Suitability for Small Systems

Some disinfection processes have already been customized for small systems. UV disinfection, in particular, is probably more appropriate for small systems that treat ground water than for large systems. UV disinfection systems require electricity to power the UV lamps in the device. If water is pumped during treatment, the UV device could be wired to operate whenever the pump runs. This sort of arrangement lends itself to operation without an operator in attendance, although some monitoring is needed to verify that the UV disinfection process is operating properly when water is being pumped.

A key factor related to the use of free chlorine in small systems is feeding the chlorine. A number of chemical solution feeders are available for feeding calcium hypochlorite or sodium hypochlorite solutions. Sodium hypochlorite is easily added to the water using a diaphragm pump, but calcium hypochlorite sometimes contains insoluble particles that can cause problems with these solution feeders. To prevent such problems, some types of equipment can add hypochlorite to the water in solid form. One such feeder discharges small hypochlorite pellets at a measured rate that can be changed by adjusting the feeder. This chlorinator typically is mounted near the top of a well casing and wired to operate whenever the well pump runs. This way, hypochlorite pellets drop into the well casing whenever water is pumped. This type of feeder is most appropriate for disinfection of ground water, but a clever operator probably could adapt it to treatment of surface waters. Another type of chlorine feeder works on the erosion feed principle. In this device, hypochlorite disks shaped like hockey pucks slowly dissolve when water flows through the feeder. This feeder has the advantage of being able to operate without electrical power, but a disadvantage is the fluctuation of chlorine dose that results from uneven rates of dissolution of the hypochlorite disks. A possible solution to the problem of uneven feed rates would be use of an equalization tank ahead of the tank or basin providing chlorine contact time. The equalization tank would be designed to dampen fluctuations in influent chlorine concentration and provide a more steady effluent chlorine concentration.

A number of manufacturers make small package ozone-generating systems. To use ozone, a utility must also provide a contactor or series of contactors. Typically these need to be 6 m (20 ft) deep to provide for efficient contact between ozone and the water being treated as the ozone bubbles added at the bottom rise to the top of the contactor. Some small systems might have flows low enough that ozone contact chambers could be made from large-diameter reinforced concrete pipes placed in the ground and aligned on a vertical axis. A number of ozone systems use ejector or diffuser systems that do not require deep contact basins.

Within the present regulatory framework, for most small systems use of free chlorine will be the easiest disinfection process to manage because of the greater complexity associated with using the other disinfectants. If free chlorine causes formation of disinfection byproducts, a logical next step would be to use free chlorine for pathogen kill and chloramine to provide a distribution system residual. If that approach does not adequately control disinfection byproducts, use of ozone followed by chloramine would be appropriate.

Corrosion Control

How the Process Works

Many water systems include corrosion control technologies to prevent corrosion of the water distribution system and to reduce lead and copper concentrations in the water where lead and copper pipes or fittings are used. Corrosion control generally involves modifying the chemistry of the water, forming a precipitate or stabilizing compound on the surfaces of piping in contact with water, or both. Most approaches include adding chemicals that can increase the alkalinity or pH of the water or can act as corrosion inhibitors by lining pipe surfaces.

One approach to corrosion control for small systems is the use of limestone contactors to modify water chemistry. Instead of using a feeder to add chemicals that increase alkalinity and pH, low-pH, corrosive water is passed through a bed of limestone rock. The water dissolves the calcium carbonate in the limestone, increasing the alkalinity and pH. One advantage of this approach is that because the chemicals are added to the water by dissolution, they cannot be overdosed, as could happen during a malfunction of a chemical feeder. Letterman et al. (1987) have shown that this process can work for small water systems, and the application of a limestone contactor for a small water system was discussed by Benjamin et al. (1992). An approach for steady-state design of limestone contactors was described by Letterman et al. (1991).

Another approach to corrosion control is the use of orthophosphates and polyphosphates (AWWARF, 1996). Orthophosphates are effective corrosion inhibitors at concentrations of 1 to 3 mg/liter as phosphate. They will aid in the reduction of lead and copper concentrations at the tap and will also reduce the rate of iron corrosion. Polyphosphates are effective as an agent to prevent red water, an undesirable effect of iron corrosion, because they will complex the iron before it can form a reddish precipitate. They also revert to orthophosphates, and this is thought to be a major reason for their effectiveness in controlling lead and copper concentrations at the tap.

Some ground waters have high concentrations of carbon dioxide (CO_2). For such waters the removal of CO_2 by air stripping can raise the pH and reduce corrosivity. Air stripping is especially useful for copper corrosion control in low-pH, high-alkalinity waters (Edwards et al., 1996).

Appropriate Water Quality and Performance Capabilities

Chemicals added through feeders can change the pH of water to virtually any desired value, depending on the type and concentration of corrosion control chemical being fed.

The range of pH and alkalinity increase that can be attained by limestone contactors is limited by the equilibrium chemistry for calcium carbonate solubility. Thus limestone contactors have a practical upper limit for the pH of treated water. If a high-pH approach to corrosion control is desired, limestone contactors will not suffice. In addition, waters containing reduced, dissolved iron could cause problems in a limestone contactor if the pH increase is sufficient to precipitate iron onto the limestone rock in the contactor. Turbidity also might foul a contactor. For these reasons, the quality of the water to be treated by a limestone contactor should be evaluated before a contactor is installed.

For orthophosphates and polyphosphates, pH control is important, because the orthophosphates work best at a pH in the range of 7.2 to 7.8 for lead and copper control.

Aeration to strip CO_2 from ground water could result in oxidation of dissolved iron and thus might be inappropriate for some waters or might require use of additional treatment processes for removal of precipitated iron.

Monitoring and Operating Requirements

Distribution system and customer tap monitoring requirements for corrosion control are set forth in the EPA's Lead and Copper Rule. In addition, corrosion control process equipment should be monitored as a means of maintaining control of the treatment process. Chemical feeders require regular checking for operational status, feed rate, and amount of chemical fed during the time interval since the last check. Limestone contactors should be inspected periodically to determine the amount of limestone remaining in the contactor. (Because limestone dissolves, it must be periodically replaced.) Regular inspections to check for fouling are also wise.

Suitability for Small Systems

Chemical feeders for use in small water systems are readily available, but determining and adjusting chemical feed rates may be difficult for small systems. Water quality problems can result from both underfeeding and overfeeding pH adjustment chemicals or corrosion inhibitors. The dosages must be correct for the corrosion control chemicals to work properly, so careful monitoring is required. In contrast, the limestone contactor concept for corrosion control was developed specifically for small systems, and if raw water quality is amenable to this treatment technique it is well suited to small systems. Use of aeration for stripping CO_2 from ground water also is a manageable process for small systems, although it must be carefully controlled to prevent excessive calcium carbonate precipitation in the distribution system.

Membrane Filtration Systems

How the Process Works

Once considered a viable technology only for desalination, membranes are increasingly employed for removal of bacteria and other microorganisms, particulate material, and natural organic matter, which can impart color, tastes, and odors to the water and react with disinfectants to form disinfection byproducts. As advancements are made in membrane production and module design, capital and operating costs continue to decline.

The several membrane filtration technologies appropriate for water treatment are distinguished by their nominal pore size or nominal molecular weight cutoff (MWCO). The MWCO is an estimate of the smallest size molecule that will be retained by the membrane in a filtration process. By these guidelines, membrane filtration technologies are classified as employing microfiltration, ultrafiltration, or nanofiltration, with microfiltration using the largest pores and having the highest MWCO and nanofiltration using the smallest pores and having the lowest MWCO (see Figure 3-1). All three types use similar principles.

Pressure-driven membrane filtration systems use applied pressure to drive water from the source water side of a semipermeable membrane to the produced-water side. Impurities are retained by size separation on the membrane while the water passes through the membrane, and they concentrate in the retained concentrate stream. The membrane permeate or product water is generally of a very high quality.

Membranes are thin, porous structures produced from a variety of materials. Early membranes were commonly made of cellulose acetate, and this type of membrane remains a choice today. Membranes are also now made of polypropylene, polyethylene, aromatic polyamides, polysulfone, and other polymers. Each membrane material has relative advantages and disadvantages. Cellulose acetate membranes permit fairly high water flux but are limited to operation in fairly narrow ranges of temperature (less than 30°C) and pH (3 to 6) and are sensitive to chlorine. Polyamide membranes have a higher resistance to pH and temperature extremes but are similarly intolerant of chlorine. Polysulfone membrane materials are more resistant than either of the other types to pH extremes, temperature, and chlorine exposure but, being hydrophobic, may foul more rapidly. Reliable, durable membranes are presently available, but the science of membrane produc-

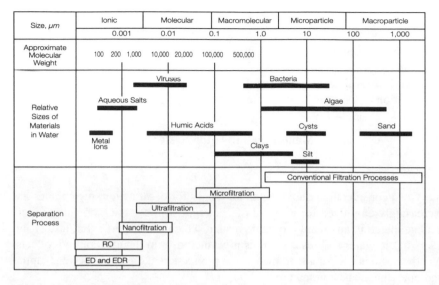

FIGURE 3-1 Sizes of molecules removed by various membrane processes in comparison to conventional filtration processes. SOURCE: Reprinted from *Electrodialysis and Electrodialysis Reversal* (M38), by permission. ©1995 by the American Water Works Association.

tion is still advancing. In addition, while all membrane materials work well under the proper conditions, choosing the most appropriate membrane for a given application still remains an art. The longevity of various membranes should be compared based on manufacturer information prior to choosing a given membrane material.

Many membranes are anisotropic in nature, consisting of a thin surface skin approximately 0.1 to 2 microns in thickness supported by a sturdier, more porous structure 100 to 200 microns in thickness (Cheryan, 1986). The surface skin performs the needed sieving of impurities from water. Composite membranes are also available. These consist of a highly resistant porous polymer, such as polysulfone, coated with a highly selective skin layer, such as cellulose acetate. Membranes can also be surface treated, as in a surface-sulfonated polysulfone membrane. This modified surface is more hydrophilic than the parent polymer, thus reducing fouling potential.

Membranes can be arranged in any of several types of configurations, the most common being hollow fine-fiber modules and spiral-wound modules. In either setup, the operating principle is the same. Water is pushed through the membrane by a higher upstream pressure. Contaminants are removed from the permeate water by sieving. Hollow fine-fiber membrane modules consist of thousands of hollow membrane tubes, approximately 500 to 1,000 microns in

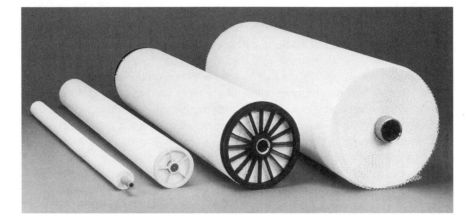

FIGURE 3-2   Spiral-wound membrane elements. SOURCE: Courtesy of Osmonics, Minnetonka, Minnesota.

diameter each, with the selective skin layer either on the interior or exterior surface of the tube. If the skin is on the interior surface of the tubes, pressurized source water is fed through the inside of the tubes, permeate water passes through the pores in the membrane, and the concentrate water with its impurities remains inside the fibers. The concentrate water flows out the opposite end of the membrane tubes and can be sent through a series of membrane modules for further treatment. An advantage of hollow fiber modules is the low pressure drop within a membrane module in comparison to spiral-wound modules, meaning that power requirements are lower for these units than for spiral-wound modules.

A spiral-wound module is made up of multiple sheets of flat membranes, with a mesh spacer material sandwiched between (see Figure 3-2). In order to provide a large membrane surface area within a fairly small module volume, the stack of membranes is rolled like a jelly roll, with the influent water fed to the individual membrane sets by a tube in the center of the roll, hence the term "spiral-wound." The membranes are arranged in sets of two, with the selective surfaces of the two membranes facing each other in each set. Source water passes under pressure through the interior surface of each set. Permeate water passes through the membranes and collects in the channels between membrane sets, then flows to a permeate water collection system. Concentrate water remains in the channel within the membrane sets, and, as with the hollow fine-fiber modules, this water can be further processed in a series of spiral-wound modules.

Appropriate Water Quality and Performance Capabilities

Membrane filtration is a physical rather than chemical treatment process.

Chemical characteristics of the water source do not greatly affect the process except in their potential for fouling the membrane surface. The concentration of particulate matter, such as bacteria and clays, and natural organic matter is of concern, for these substances can foul the membranes. To avoid this, water must be relatively free of particulate material prior to entering a membrane module. Surface waters may require pretreatment by a more conventional treatment process prior to polishing by membrane filtration, although membrane systems are capable of tolerating a lower quality surface water than direct filtration systems (discussed later in this chapter). Generally, coarse filtration, such as that provided by a bag or cartridge filter, will sufficiently pretreat the source water. Sometimes a coarser mode of membrane filtration is used prior to a finer filtration operation, such as pretreating a surface water with microfiltration prior to removal of disinfection byproduct precursors with a nanofiltration system.

Microfiltration is loosely defined as a membrane separation process using membranes with a pore size of approximately 0.03 to 10 microns, an MWCO of greater than 100,000 daltons, and a relatively low feed water operating pressure of approximately 100 to 400 kPa (15 to 60 psi). Representative materials removed by microfiltration include sand, silt, clays, *Giardia* and *Cryptosporidium* cysts, algae, and some bacterial species.

Ultrafiltration involves the pressure-driven separation of materials from water using a membrane pore size of approximately 0.002 to 0.1 microns, an MWCO of approximately 10,000 to 100,000 daltons, and an operating pressure of approximately 200 to 700 kPa (30 to 100 psi). Ultrafiltration will remove all species removed by microfiltration as well as some viruses and humic materials.

Nanofiltration membranes have a nominal pore size of approximately 0.001 microns and an MWCO of 1,000 to 10,000 daltons. Pushing water through these smaller membrane pores requires a higher operating pressure than either microfiltration or ultrafiltration. Operating pressures are usually near 600 kPa (90 psi) and can be as high as 1,000 kPa (150 psi). These systems can remove virtually all viruses and humic materials. They provide excellent protection from disinfection byproduct formation if the disinfectant residual is added after the membrane filtration step. Because nanofiltration membranes also remove alkalinity, the product water can be corrosive, and measures such as blending raw water and product water or adding alkalinity may be needed to reduce corrosivity.

Membrane filtration greatly reduces the need for disinfectants. Protozoa, bacteria, and even viruses can be removed in the process, which can relieve a portion of the *CT* disinfection requirement, if proven to the satisfaction of regulators. Nanofiltration also removes hardness from water, which accounts for nanofiltration membranes sometimes being called "softening membranes." (Hard water treated by nanofiltration will need pretreatment to avoid precipitation of hardness ions on the membrane.) Although membrane filtration is most commonly used to remove inorganic or microbiological contaminants, a pilot-scale demonstration showed that a nanofiltration system removed a variety of synthetic organic chemicals (Duranceau et al., 1992). Removal was related to the molecular weight of the synthetic organic compound. Lower molecular weight synthetic organic compounds such as ethylene dibromide and dibromochloropropane passed through the membrane, while the slightly higher molecular weight pesticides chlordane, heptachlor, and methoxychlor were removed from the permeate. Based on the results of such studies, larger organic compounds such as natural organic matter would be removed by nanofiltration.

Membrane classification standards vary considerably from one filter supplier to the next. One supplier may sell as an ultrafiltration membrane a product similar to what another manufacturer calls a nanofiltration system. It is best to look directly at pore size, MWCO, and applied pressure needed when comparing two membrane systems.

Monitoring and Operating Requirements

Efficient operation of a membrane separation system relies as much on module design as on membrane material choice. Capital costs of membrane systems are a function of the type of system configuration and the membrane surface area: volume ratio for a given module. Operating costs are influenced by module replacement costs, pressure requirements, ease of cleaning, and cleaning solution and concentrate disposal costs. While the initial membrane purchase is a relatively minor portion of the capital cost, membrane replacement is the largest component in the cost of operation (Wiesner et al., 1994).

Prevention of fouling of microfiltration and ultrafiltration membranes requires regular backwashing of the membranes. Operation is usually automated, with backwash of contaminants from the membrane surface occurring at a prearranged time, a prescribed effluent turbidity, or a predetermined change in operating pressure. For this reason membrane plants often can be allowed to operate unattended much of the time. The principle of operation is simple and not tied directly to source water chemistry. Antiscalant chemicals may need to be added to the water when the concentrated water retained by the membrane exceeds solubility limits for salts such as calcium carbonate. This is more likely in tighter membrane systems such as those using nanofiltration.

Waste stream disposal is a significant problem in many areas. Unlike conventional treatment processes, in which approximately 5 to 10 percent of the influent water is discharged as waste, membrane processes produce waste streams amounting to as much as 15 percent of the total treated water volume. Because little or no chemical treatment is used in a membrane system, the concentrate stream usually contains only the contaminants found in the source water (although at much higher concentrations), and for this reason the concentrate can sometimes be disposed of in the source water. Other alternatives include deep well injection, dilution and spray irrigation, or disposal to the municipal sewer; these alternatives are usually necessary for nanofiltration waste, which usually contains concentrated organic and inorganic compounds. Regardless of the type of membrane, concentrate disposal must be carefully considered in decisions about the use of membrane technology.

Suitability for Small Systems

Membrane filtration systems have very little economy of scale, so capital costs on a basis of dollars per volume of installed treatment capacity do not escalate rapidly as plant size decreases. This makes membranes quite attractive for small systems. In addition, for ground water sources not needing pretreatment, membrane technologies are relatively simple to install, and the systems require little more than a feed pump, a cleaning pump, the membrane modules, and some holding tanks. Most experts expect that membrane filtration will be used with greater frequency in small systems as the complexity of conventional treatment processes for small systems increases.

In a cost comparison of membrane filtration and conventional treatment, particle removal by ultrafiltration was estimated to be substantially less expensive than by conventional filtration technologies for small systems (Wiesner et al., 1994). As facility capacity decreased, the membrane cost advantage increased. Similarly, when nanofiltration was compared to conventional treatment with the addition of ozone and granular activated carbon to control disinfection byproduct and total organic carbon levels, the two treatment techniques produced similar water quality, but the membrane systems were substantially less costly for small system sizes (Wiesner et al., 1994).

The operation of a nanofiltration system is substantially less complicated than operation of the multiple treatment train needed to reach the same result by conventional systems. Membrane filtration should be considered for small systems that need to remove multiple contaminants. There are few limitations to the types of raw water that membrane filtration systems can treat, although pretreatment of the water to remove particles may be necessary, and testing to determine potential fouling by organic matter should be performed.

Reverse Osmosis

How the Process Works

Reverse osmosis (see Figure 3-3) is a highly efficient removal process for inorganic ions, salts, some organic compounds, and, in some designs, microbiological contaminants. Reverse osmosis resembles membrane filtration processes in that it involves the application of a high feed water pressure to force water through a semipermeable membrane. In osmotic processes, water spontaneously passes through a semipermeable membrane from a dilute solution to a concentrated solution in order to equilibrate concentrations. Reverse osmosis is pro-

FIGURE 3-3 Skid-mounted reverse osmosis system. SOURCE: Courtesy of Osmonics, Minnetonka, Minnesota.

duced by exerting enough pressure on a concentrated solution to reverse this flow and push the water from the concentrated solution to the more dilute one. The result is a clear permeate water and a brackish reject concentrate.

Several differences distinguish reverse osmosis from membrane filtration. Unlike in membrane filtration, in reverse osmosis the membrane is essentially nonporous; transport of water through the membrane takes place by sequential dissolution of the water into the membrane and diffusion through the membrane to the permeate side. Any contaminants that can dissolve into and diffuse through the membrane can also pass into the permeate in this system, though such contaminants are few. The membrane rejects most solute ions and molecules, allowing water of very low mineral content to permeate; some organic contaminants can permeate reverse osmosis membranes.

Reverse osmosis produces a larger volume of reject concentrate solution than membrane filtration. The concentrate volume can be as much as 25 to 50 percent of the raw water volume. In addition, though module configurations resemble those of membrane filtration processes, operating pressures are much higher, ranging from approximately 1,400 kPa (200 psi) for water with a total dissolved solids concentration of less than 1,000 mg/liter to as high as 10,000 kPa (1,500 psi) for seawater with a total dissolved solids content of 35,000 mg/liter. The higher pressure is needed to overcome the solution osmotic pressure and force water through the membrane from the concentrated feed side to the dilute permeate side.

The permeate from a reverse osmosis system is virtually demineralized and therefore quite corrosive. To maintain stable water in the distribution system, a predetermined fraction of the raw water is usually allowed to bypass the system and is mixed with the permeate. Posttreatment may include degasification if carbon dioxide and/or sulfide is present in the water, pH adjustment to reduce corrosiveness, and disinfection.

Appropriate Water Quality and Performance Capabilities

Removal efficiencies for inorganic ions and salts range from 85 to 99 percent. Removal of organic chemicals varies with the chemical in question. Low-molecular-weight organic compounds, as well as organic compounds with an affinity for the particular membrane material, may diffuse through the membrane. The removal efficiencies for organic compounds range from no removal to better than 99 percent removal. Humic materials, particulate matter, microorganisms, and viruses are generally removed in the process, but the bypass water will add microbiological contaminants to the treated water when the two are mixed to reduce corrosiveness. Also, leaks of concentrate water containing bacteria and viruses can occur around o-ring seals under the high operating pressures of a reverse osmosis system.

A reverse osmosis membrane can severely foul if proper pretreatment is not provided. Influent total organic carbon concentrations can be as high as 20 mg/liter, but pretreatment must be used to reduce influent turbidity and to remove any iron, manganese, and chlorine. Stabilization of the water to prevent scale formation may also be necessary, as the concentrate solution may contain inorganic contaminant concentrations so high that precipitation could occur. The water's pH may have to be adjusted to avoid reducing membrane life.

A typical membrane module lasts 3 to 5 years, after which module replacement is necessary. Membrane module replacement costs remain high and are a significant consideration in the overall cost of the treatment system.

Monitoring and Operating Requirements

Most reverse osmosis systems are set up to backwash automatically, and therefore the pressure unit itself requires little operator attention. However, pretreatment may require a skilled operator.

The membrane must be cleaned periodically to remove scale at the surface. Caution is required to avoid contamination of raw or finished water with the generally acidic cleaning solution. In addition, membranes must be flushed with product water prior to shutdown to prevent prolonged contact between the membrane and a concentrated solution; otherwise, scaling from chemical precipitation can occur within hours of shutdown. If the plant is not operated for several days, the membranes should be filled with a disinfectant solution to prevent biological growth and possible membrane damage.

Disposal of reject water poses an even greater problem for reverse osmosis systems than for lower-pressure membrane filtration processes such as ultrafiltration or nanofiltration because they produce a larger quantity of reject water, and the contaminants are more concentrated than those produced by filtration processes. Release to evaporation ponds or the municipal sewer or injection in deep wells are current disposal strategies. However, in the future some of these strategies may no longer be permitted in some areas. Disposal needs and local regulations governing disposal must be considered in planning a reverse osmosis treatment plant.

Suitability for Small Systems

Like other modular membrane processes, reverse osmosis has little economy of scale and therefore is just as suitable for a small system as it is for a large one. Reverse osmosis is a rugged and reliable treatment process on the small scale. Air-droppable reverse osmosis units were used during the Gulf War to supply water to troops near saline water supplies. Plant expansion can be as easy as adding an additional series of membrane modules to the treatment train. Operation can be automated, allowing reverse osmosis systems to be run by part-time operators. There are 142 operating reverse osmosis drinking water plants in North America, with more than a third of them serving fewer than 3,500 people (Morin, 1994). The technology is commonly used in Florida to treat drinking water for condominiums and mobile home parks (Sorg et al., 1980).

One example of a small community that uses reverse osmosis is Wenona, Illinois. Prior to installation of a reverse osmosis system, the town's approximately 1,200 residents experienced problems with the deep-well ground water they use. The source water has high levels of dissolved solids, which imparted a salty taste to the water and damaged equipment such as water heaters and washing machines. Most residents drank bottled water. In addition, radium levels in the source water are above the drinking water standard. The reverse osmosis plant removes 99 percent of radium-226 and 95 percent of dissolved solids. Most consumers have since taken their water softeners off line and are able to use approximately half of the soap and shampoo once necessary (JAWWA, 1993).

Electrodialysis/Electrodialysis Reversal

How the Process Works

Electrodialysis (ED) and electrodialysis reversal (EDR) systems, usually employed to produce demineralized water from brackish water sources, use electrochemical separation processes to concentrate salts from the feed water into a smaller-volume, higher-concentration solution. ED and EDR systems consist of stacks of alternating anionic and cationic selective membranes. The ionic components of dissolved salts pass through the membranes in response to an electric current applied to the water perpendicularly to the membranes. The system creates a demineralized product water stream and a brine concentrate stream.

In ED and EDR systems, the anions travel from the feed water channel toward the anode and pass through an anionic selective membrane but are rejected from transfer through the cationic selective membrane; the result is that anions are retained in the channel between the anionic and cationic membranes (see Figures 3-4 and 3-5). Simultaneously, cations from another feed channel travel toward the cathode in response to the electric current, pass through a single cationic membrane, and are concentrated in the same channel as the anions between the cationic and anionic membranes. In this manner, all the ions in a given feed channel are removed and concentrated in a concentrate channel.

In EDR, the polarity of the electrodes is reversed every 15 to 20 minutes. The change causes a reversal in ion movement. A concentrate channel at one polarity becomes a demineralized channel at the opposing polarity. Automatically operated valves tied in to the polarity change transfer incoming and outgoing flows to the proper piping. Reversing the polarity, and consequently the water flows, minimizes scale buildup by providing regular washing of the membrane surface. EDR systems can thus operate for longer periods of time between cleanings than ED systems. The majority of plants in the United States using this technology are EDR plants (Morin, 1994).

Appropriate Water Quality and Performance Capabilities

ED and EDR systems require feed water pretreatment, at a minimum with cartridge filters. Feed water turbidity should be less than 2.0 nephelometric turbidity units (NTU), the free chlorine concentration less than 0.5 mg/liter, manganese less than 0.3 mg/liter, and hydrogen sulfide less than 0.3 mg/liter (Conlon, 1990). Hydrogen sulfide is highly unlikely to be present in surface water and would generally be a concern only for ground water sources. Total dissolved solids levels of up to 4,000 mg/liter have been tolerated by EDR plants successfully producing water that meets drinking water total dissolved solids standards (Morin, 1994).

In contrast to membrane filtration processes and reverse osmosis units, the product water in ED and EDR systems does not pass through the membrane. This reduces the potential for concentration polarization and organic fouling of the membrane surface but provides no means for removing microbiological contaminants, organic compounds, or particulate or colloidal materials. SDWA requirements for these contaminants must be met through pretreatment or posttreatment of the water by other means, if necessary.

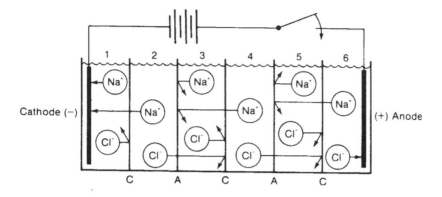

FIGURE 3-4 Removal of sodium chloride from water in an ED system. Chloride ions move toward the anode and pass through the anion selective membranes (A) but are trapped by the cation selective membranes (C); sodium ions move toward the cathode and pass through the cation selective membranes but are trapped by the anion selective membranes. The result is a demineralized water in channels 2 and 4. SOURCE: Reprinted, with permission, from Meller (1984), *Electrodialysis (ED) and Electrodialysis Reversal (EDR) Technology.* ©1984 by Ionics, Inc., Watertown, Massachusetts.

FIGURE 3-5 Typical membranes and membrane spacers used in ED and EDR systems. SOURCE: Courtesy of Ionics, Inc., Watertown, Massachusetts.

Both the anionic and cationic membranes used in ED and EDR systems are fairly sturdy and resistant to water quality conditions. They are 0.5 mm in thickness and resist damage from pH extremes in the range of pH 1 to 10. They can tolerate temperatures up to 46°C (115°F) (Conlon, 1990).

A disadvantage of ED and EDR systems is their high energy requirements. Pumping requirements are similar to those for an ultrafiltration system, with the costs for maintaining the direct current at least equal to the pumping costs. Energy costs other than pumping are a function of feed water salinity. The energy required to provide the current is approximately 2.0 kWh per 3,800 liters (1,000 gals.) treated per 1,000 mg/liter total dissolved solids removed (Conlon, 1990).

The recovery rate in an ED or EDR process is the percentage of feed water that becomes product water. Most EDR plants operate at a recovery rate of 70 percent or better. The remaining 30 percent is disposed of as the concentrate stream (Morin, 1994). Like water treated by reverse osmosis and nanofiltration, ED/EDR water is corrosive, so some bypassed water may be needed to stabilize the product water.

Monitoring and Operating Requirements

ED and EDR systems are usually fully automatic. Recordings of system operation may be taken by a computer or by an operator, if available. Membranes and, less frequently, electrodes will need to be replaced. Routine maintenance is fairly simple. Equipment such as pumps and chemical feed systems requires the usual maintenance. ED (but not EDR) systems need antiscalant chemicals.

Operation must be performed at a direct current density less than the limiting density of the system. A limiting voltage also applies in order to prevent heating the system and causing damage to the membranes and/or spacer material. These operating parameters are set at installation by the supplier.

Suitability for Small Systems

ED and EDR plants are well suited to small systems with brackish water sources. More than half of the operating ED and EDR plants in North America serve fewer than 3,500 people (Morin, 1994). Some plants serve as few as 200 people. ED and EDR plants are generally automated, allowing for part-time operation. As with reverse osmosis, energy consumption must be considered when evaluating whether to apply this technology.

Adsorption

How the Process Works

Adsorption is the physical and chemical process by which an organic contaminant accumulates on the surface of a solid, removing the contaminant from solution in the water. Organic contaminants, including toxic synthetic organic chemicals, color-causing compounds, and taste- and odor-causing compounds, are all less polar than water and therefore have low solubility in water, which is a polar liquid. Thus, they are attracted to the nonpolar solid surface.

The most common adsorbent used to remove organic contaminants from water is activated carbon. Activated carbon is similar to charcoal. It differs from charcoal in that the base material (typically bituminous coal, lignite, petroleum coke, or bone char) has been heated in the absence of air to carbonize it and then activated by oxidation at 200°C to 1,000°C to develop a favorable pore structure. The result is a highly porous structure with a very high surface area per unit volume, which allows for significant adsorption of impurities from water. In general, the less soluble an organic compound, the better it adsorbs from solution onto the activated carbon (Lundelius, 1920; Weber, 1972).

Activated carbon is available in two common forms, powdered activated carbon (PAC) and granular activated carbon (GAC), the difference between the two being obvious from their names. PAC is generally less than 50 μm in diameter and is added to the raw water line or to a mixing basin. For effective treatment, the PAC must contact all of the incoming water. Because of its small particle size, adsorption to the surface occurs quickly. The normal contact time of mixing basins used for other elements of water treatment is sometimes sufficient for contaminant adsorption onto PAC. In such cases, no modification other than the addition of PAC dosing equipment needs to be made to an existing plant. In other cases, adsorption can require up to 8 hours of contact time. Testing prior to design is needed to determine the required contact time. Following adsorption, the carbon containing the organic compounds is settled or filtered from the water and disposed of with the plant sludge.

GAC has a grain size in the range of 0.5 to 1.5 mm, 10 to 100 times larger than PAC. It is packed into columns through which the raw water flows. Packing the carbon in columns allows more complete contact between the water and activated carbon, greater adsorption efficiency, and greater process control than is possible with PAC (Snoeyink, 1990). GAC can be removed from the column for carbon regeneration or reactivation when necessary.

In addition to being used to adsorb organic compounds, GAC systems are sometimes used as biological filters. Microbes that stabilize water quality are allowed to grow on the GAC surfaces and in particle filters.

Appropriate Water Quality and Performance Capabilities

Activated carbon adsorption historically was used primarily to remove tastes and odors from water, but its use as an adsorbent for toxic or carcinogenic organic compounds has increased steadily and is now a primary application.

PAC should be added prior to filtration in order to provide for removal of all the powdered material. PAC can be very economical if it is only needed on a periodic basis in response to changes in influent water quality. The dose of PAC added to the system can also be adapted to deal with varying source water quality.

GAC use requires removing particulate material from the untreated water to avoid clogging the treatment column. An alternative is to use the GAC column as the filter medium, performing both filtration and adsorption in a single step. This method requires frequent backwashing of the carbon column. Backwashing mixes the carbon in the column and can cause spent carbon to be deposited near the column effluent. The spent carbon might release some of the target compound to the effluent water. If this can be tolerated, the method is simple and compact.

Competitive adsorption is an important consideration in the design of an activated carbon system for drinking water. Natural organic material in the water may compete with contaminants for adsorption sites on the carbon, increasing the amount of carbon needed to remove the target contaminant. Competing organic compounds can also displace contaminants already adsorbed to the carbon. If the competing compound is present in a high concentration, it may displace the adsorbed contaminants to such an extent that the effluent concentration of the contaminant is temporarily greater than the influent concentration. For this reason, competing chemicals must be removed from the system prior to adsorption, or the system design and carbon replacement frequency must be adequate to allow for the competition.

Monitoring and Operating Requirements

In PAC systems, care must be taken to remove all PAC from the water. This usually requires filtration. Even a small amount of PAC passing through the system can cause the water to turn gray. In addition, if PAC enters a sample vial used for determining whether the treated water meets drinking water standards, the apparent aqueous concentration of the target contaminants can exceed regulatory standards because these contaminants will be concentrated on the PAC.

Single-stage GAC systems (meaning those in which the water flows through one GAC column rather than two or more in series) must be monitored to ensure that the column is taken out of service as soon as any trace of the target compound is found in the effluent water. If an exhausted GAC column is not regenerated or replaced, no adsorption will occur, and desorption may result in the effluent having higher concentrations of some contaminants than the influent. In addition, a buildup of microorganisms in the column may clog the column or create taste and odor problems. Monitoring for organic compounds is not as simple as for some inorganic contaminants and will likely require the services of an experienced operator or outside laboratory to perform the analyses. In either case, these monitoring requirements will increase the cost of GAC implementation.

Suitability for Small Systems

GAC is quite easy to apply on a small scale; small columns can be readily obtained and installed. Virgin (rather than regenerated) carbon is often required for use in drinking water applications, which can increase the operating costs. PAC is also quite simple to employ on a small scale if the plant already uses a process train including mixing, precipitation or sedimentation, and filtration.

Lime Softening

How the Process Works

In the lime-softening process, the pH of the water being treated is raised sufficiently to precipitate calcium carbonate and, if necessary, magnesium hydroxide. Calcium and magnesium ions in water cause hardness; hard water can cause scaling problems in water heaters, and soap lathers poorly in hard water. Therefore, some water utilities remove calcium and magnesium to soften the water and improve its quality for domestic use. In small systems, lime softening would typically be practiced by adding hydrated lime to raw water to raise the pH to approximately 10. This removes calcium carbonate. If magnesium removal is also required, the pH during softening would need to be closer to 11. In some waters, addition of soda ash is needed for effective hardness removal. After mixing, flocculation, sedimentation, and pH readjustment, the softened water is filtered.

Appropriate Water Quality and Performance Capabilities

Many large water systems in the midwestern United States use lime softening to treat surface waters from sources such as the Missouri and Mississippi rivers. Well-operated lime softening plants can cope with a range of quality as great as that treated by conventional treatment. However, the combination of variable source water quality and the complexity of the chemistry of lime softening may make lime softening too complex for small systems that use surface water sources. Lime softening may be more appropriate for small systems that use ground water because of the relatively uniform quality of ground water. Once the softening chemistry for a ground water is determined, it should not change much. In comparison, chemical additions to surface waters need to be modified frequently in response to water quality changes.

In addition to removing calcium and magnesium, lime softening removes radium, which is chemically similar to calcium and magnesium. It also removes arsenic, oxidized iron, and manganese. A recent study (Logsdon et al., 1994) indicated that lime softening plants may remove *Giardia* cysts as effectively as conventional treatment plants.

Monitoring and Operating Requirements

Regulatory monitoring requirements for lime softening plants depend on whether the source water is surface water or ground water. Process monitoring requirements should focus on measurement of pH, hardness, and alkalinity for plants treating ground water. In addition, filtered water turbidity monitoring is needed at plants treating surface water, not merely for compliance purposes but also to manage filter operation.

One of the difficult aspects of lime softening is the operation and maintenance of lime feeders and lines carrying lime slurry to the point of application. In addition, plant operators must understand lime softening chemistry. Measurement of pH must be accurate, and the operator must know that the pH meter is properly calibrated. Failure to maintain the proper pH in softened water prior to filtration at a lime softening plant could result in precipitation of excess lime in the filter beds and formation of calcium carbonate (essentially limestone) deposits within the filters. Because of these operational difficulties, in the future, small systems that decide to soften water may seriously consider using nanofiltration or reverse osmosis for softening instead of chemical precipitation.

Suitability for Small Systems

Lime softening has been used successfully by ground water systems serving fewer than 3,000 people. Lime softening is not likely to be applied with success by small systems that treat surface waters because of the complexity of the chemistry involved. In addition, lime softening is unlikely to be suitable for treating ground water in systems serving 500 or fewer people unless those systems have some form of contract or satellite operation that would enable a trained operator to monitor the treatment process periodically. Prefabricated lime softening equipment is available for use by small systems.

TECHNOLOGIES FOR SYSTEMS WITH
GROUND WATER SOURCES

Ground water sources generally require less treatment than surface water supplies or ground water supplies that are under the influence of surface water. Natural filtration in the subsurface reduces the concentration of many substances, including those that cause turbidity.

Ground water has generally been considered free of microbiological contamination, and throughout much of the twentieth century many ground water supplies have been distributed without treatment. Ground water has been implicated in some disease outbreaks, however (Macler, 1996). As a result, the EPA has issued a proposed Ground Water Disinfection Rule (GWDR) that would require disinfection of all ground waters except those that qualify for a variance or meet "natural disinfection" requirements as determined from an evaluation of criteria such as setback distance from potential contamination sources, ground water travel time, and local hydrogeologic features (Grubbs and Pontius, 1992). All systems would be required to maintain a detectable disinfectant residual in the distribution system at all times or to maintain a heterotrophic plate count level of less than 500 organisms per milliliter. Grab samples of disinfectant would be required one to four times per day under the proposed rule.

Many ground water supplies have been contaminated from improperly sealed wells, septic tank effluent, chemicals from agricultural use, leaking underground storage tanks, and leachate from waste disposal sites. In coastal areas, overpumping of ground water has led to saltwater intrusion. These newer forms of contamination have made it necessary to treat many ground water supplies prior to disinfection and distribution. As mentioned in Chapter 2, the most common chemical contaminants in ground water are nitrate, fluoride, and volatile organic chemicals.

In addition to the technologies mentioned above, which are suitable to both surface and ground water systems, some technologies are best suited to the types of contamination found in ground water. These include air stripping, oxidation/ filtration, ion exchange, and activated alumina. These processes (and others) are outlined in Table 3-3 along with the contaminants they address, the state of the technology, and a relative estimate of capital costs. Table 3-4 lists operating characteristics of the technologies.

Air Stripping

How the Process Works

Air stripping, also commonly called aeration, involves continuous contact of air with water to allow aqueous contaminants to transfer from water into the air. The air is swept from the system, taking contaminants such as volatile organic chemicals, taste- and odor- causing compounds, and radon gas out of the water, or reducing the carbon dioxide concentration to raise the pH. The contaminated air is then treated if necessary and released to the atmosphere. The driving force for transfer of the contaminants is the difference between the concentration of the contaminant in the untreated water and the concentration in water that is at equilibrium with the air. An air stripping system can remove concentrations of contaminants of up to several parts per million.

Appropriate Water Quality and Performance Capabilities

Air stripping equipment must provide for a large area of contact between the air and water and for convective movement of the water or air to allow as much water as possible to contact air. This can be accomplished in several ways:

through diffused aeration, mechanical aeration, packed tower air stripping, or gas-permeable membrane air stripping.

Diffused aeration involves introducing compressed air into the bottom of a water basin through a series of diffusers. Although usually low in cost and easy to operate and maintain, this aeration mechanism does not provide for convective movement of the water and thus does not allow as much contact between the air and water as other methods. Because of its limited efficiency, it is generally used only to adapt existing plant equipment.

Mechanical aeration introduces air into the water by rapidly agitating the water surface with a mechanical mixer. Like diffused aeration, this is not a very efficient contacting system. Mechanical aerators often require large basins, long residence times, and high energy inputs. Because they are adaptable to existing basins, mechanical aerators are often installed as a system retrofit rather than as a new design.

Tray aerators (see Figure 3-6) offer an economical method of contacting air and water. As the name suggests, a tray aerator consists of a vertical series of trays down which the water flows. The water contacts air as it drips through the trays. A tray aerator can be operated with a natural draft or with a forced draft provided by an air compressor. Using a natural draft reduces operating costs but is also less efficient than using a forced draft. Slime and algae growth can pose problems with tray aerators. Though it is not particularly desirable to add additional chemicals, biological growth on the trays can be controlled by adding copper sulfate or chlorine.

Forced draft tray aerators are one step less efficient at providing contact between air and water than the next method of aeration, packed tower air stripping. In packed tower aerators, water flows down a bed of packing material such as fixed plastic grids, loose plastic rings, or loose ceramics saddles, while air flows up through the column (see Figure 3-7). The packing material breaks the water into thin sheets and droplets, creating a large and constantly changing surface area for contact between the air and water. Pretreatment for removal of microorganisms, iron, manganese, and excessive particulate matter is important for this design. Packed tower aerators have been used for decades in water and wastewater treatment. Design is currently a straightforward process for a practiced engineer.

A final method of providing contact between air and water, using gas-permeable membranes, is gaining acceptance. These systems use a membrane module made up of highly porous, small-diameter, hollow fiber membranes as a contacting device, providing an air-water contact area per equipment volume nearly an order of magnitude greater than packed tower air strippers (Semmens et al., 1989; Zander et al., 1989). Water flows through the interior of hollow membrane tubes constructed of a material that allows gases but not liquids to pass through. The large surface area for air-water contact allows for removal of semivolatile as well as volatile organic chemicals. Gas-permeable membrane systems offer the high-

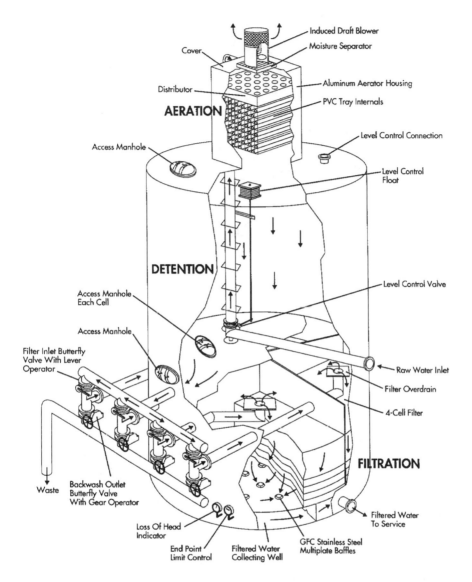

FIGURE 3-6   Cutaway diagram of a tray aerator in a package treatment system.
SOURCE: Courtesy of General Filter Company, Ames, Iowa.

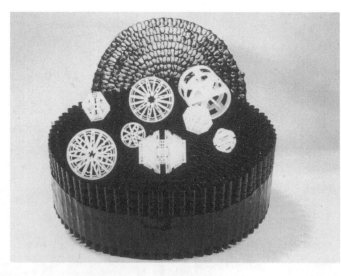

FIGURE 3-7 Typical packing material used for packed tower aerators. SOURCE: Courtesy of Delta Cooling Towers, Fairfield, New Jersey.

est removal efficiencies of all contacting devices. However, the technology must be considered "emerging" because long-term performance has not been evaluated.

Monitoring and Operating Requirements

Air stripping systems are generally set for automatic operation. They usually require only daily visits to ensure proper equipment operation and to provide preventive maintenance. Remote monitoring of pumps can indicate system performance and further reduce the need for operator attention.

Pretreatment of water may be necessary in order to avoid fouling the systems with microbial growth (especially iron-oxidizing bacteria), particulate matter, calcium precipitates, or iron precipitates. Reduced iron or manganese in ground water will oxidize when exposed to air and will precipitate. High concentrations of these chemicals can completely plug an air stripping system if not removed prior to this treatment process.

Suitability for Small Systems

An aeration system can generally be installed for a fairly low cost. The treatment process is highly adaptable to small treatment plants, often involving a simple retrofit to existing treatment basins. Cost and treatment efficiency both increase with increasing system complexity. If contaminant concentrations are high and regulations require treatment of the air leaving the system, however, costs increase dramatically. If the water is hard and contains a high CO_2 concentration, air stripping will reduce the CO_2 concentration and may cause excessive precipitation of calcium carbonate. Other than the presence of reduced iron in ground waters, no particular water quality issues affect the choice of aeration technology. The specific type of aerator depends only on the degree of contaminant reduction desired.

One example of a small community that uses aeration is the Blue Mountain Subdivision near Denver, Colorado, population approximately 400. The community installed a packed tower aerator to remove radon from the ground water supply. The aerator consistently removes 96 to 99 percent of the radon. An added benefit is the simultaneous removal of CO_2, which reduces corrosion problems (Tamburini and Habenicht, 1992).

Oxidation/Filtration

How the Process Works

There are multiple places in a surface water treatment train where oxidation chemicals may be used. They can be used for disinfection (discussed earlier), color removal, taste and odor control, or organic contaminant removal. However, because of the difficulty of controlling the chemistry of such reactions when water quality varies, as in surface water, it is unlikely that a small surface water system would use oxidation/filtration. The primary use of this technology by small systems is for removal of iron and manganese from ground water sources. Iron and manganese, while not primary drinking water contaminants, are responsible for many complaints in small systems. Iron or manganese spots on laundry and fixtures cause customer dissatisfaction with the water utility.

Iron and manganese are present in ground water in their reduced and very soluble forms. Before they can be removed, they must be oxidized (meaning they must lose electrons) to a state in which they can form insoluble complexes. These suspended insoluble complexes can be removed from water by filtration. Ferrous iron (Fe^{2+}) can be oxidized to ferric iron (Fe^{3+}), which more readily forms the insoluble iron hydroxide complex $Fe(OH)_3$. Reduced manganese (Mn^{2+}) can be oxidized to Mn^{4+}, which forms insoluble MnO_2. A detention time of 10 to 30 minutes following chemical addition is needed prior to filtration to allow the reaction to take place. The insoluble complexes are best removed from water using a medium with a large (>1.5 mm) effective size range in order to reduce filter head loss (Montgomery, 1985). In a similar manner, odorous sulfides (S^-) are oxidized to colloidal elemental sulfur (S_o) and removed.

Oxidation involves the transfer of electrons from the iron, manganese, or other chemicals being treated to the oxidizing agent. The most common chemical oxidants in water treatment are chlorine, potassium permanganate, and ozone. Chlorine has historically been the oxidant of choice, but its role in disinfection byproduct formation has led to questions about its use and a search for alternative oxidation strategies.

Chlorine is a strong oxidizing agent, and this, in addition to its ease of feeding and economy, are reasons for its long history of use. Chlorine is most effective at a pH below 7 due to its presence in water as hypochlorous acid under these conditions. Organic compounds should be removed prior to chlorine addition to reduce the formation of harmful byproducts.

Ozone is also a very strong oxidizing agent. Ozone produces fewer known byproducts than free chlorine on reaction with organic compounds, but ozone byproducts are still under study. Ozone is not greatly affected by pH levels in the water. As mentioned in the section on disinfection, ozone has a very short half-life and must be generated on site. This is quite energy intensive and requires an experienced operator.

Potassium permanganate is a moderately strong oxidant and is easy to feed to a system. Its addition does not cause trihalomethane formation, but possible production of other byproducts is still under study. Potassium permanganate in water produces a pink solution. If it is added in excess and not fully reduced by reacting with reduced compounds in the water supply, the resulting water will remain pink. This is not pleasing to the utility customer.

A low-cost method of providing oxidation is to use the oxygen in air as the oxidizing agent in a tray aerator. Water is simply passed down a series of porous trays to provide contact between air and water. No chemical dosing is required, which allows for unattended operation. This method is not effective for water in which the iron is complexed with humic materials or other large organic molecules. Oxygen is not a strong enough oxidizing agent to break the strong complexes formed between iron and manganese and large organic molecules.

Appropriate Water Quality and Performance Capabilities

The presence in water of other oxidizable species hinders oxidation of the desired reduced compounds. Volatile organic chemicals, other organic compounds, or taste- and odor-causing compounds may result in an oxidant demand. This additional oxidant demand must be accounted for when dosing the oxidant. Other than these possible interferences, there is no strict cutoff in water quality above which oxidation followed by filtration will not work. The expense of operation derives from the chemical use in most cases and is therefore directly related to the source water quality.

Monitoring and Operating Requirements

Oxidation followed by filtration is a relatively simple process. The source water must be monitored to determine proper oxidant dosage, and the treated water should be monitored to determine if the oxidation was successful. Filters must be backwashed. In general, manganese oxidation is more difficult than iron oxidation because the reaction rate is slower, so a longer detention is necessary prior to filtration.

Permanganate can form precipitates that cause mudball formation on filters. These are difficult to remove and compromise filter performance. In addition, doses of permanganate must be controlled carefully or remaining unreacted permanganate will lead to pink coloration of the water. If not dosed carefully, ozone can oxidize reduced manganese all the way to permanganate and result in pink water formation as well. Manganese dioxide particles, also formed by oxidation of reduced manganese, must be carefully coagulated to ensure their removal.

Suitability for Small Systems

Oxidation using chlorine or potassium permanganate is frequently applied in small ground water systems. The dosing is relatively easy, requires simple equipment, and is fairly inexpensive.

Ion Exchange

How the Process Works

Ion exchange (see Figure 3-8) involves the selective removal of charged inorganic species from water using an ion-specific resin. The surface of the ion exchange resin contains charged functional groups that hold ionic species by electrostatic attraction. As water containing undesired ions passes through a column of resin beads, charged ions on the resin surface are exchanged for the undesired species in the water. The resin, when saturated with the undesired species, is regenerated with a solution of the exchangeable ion. A large variety of synthetic resins is available for specific applications.

Generally, resins can be categorized as anion exchange or cation exchange resins. Anion exchange resins selectively remove anionic species such as nitrate (NO_3^-), carbonate (HCO_3^-), dichromate ($Cr_2O_7^{2-}$), fluoride (F^-) and the selenium-containing species selenate (SeO_4^{2-}) and selenite ($HSeO_3^{2-}$). These resins are less effective at removing chromate (CrO_4^{2-}). Anion exchange resins are often regenerated with sodium hydroxide or sodium chloride solutions, which replace the anions removed from the water with hydroxide (OH^-) or chloride (Cl^-) ions, respectively.

Cation exchange resins are used to remove undesired cations from water and exchange them for protons (H^+), sodium ions (Na^+), or, if sodium use is restricted, potassium ions (K^+). Cation exchange is often used to soften water by exchanging the calcium (Ca^{2+}) and magnesium (Mg^{2+}) for another ion, usually

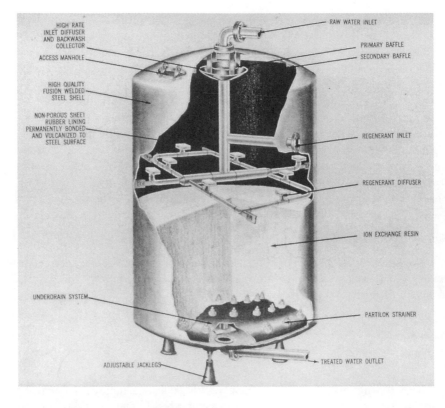

FIGURE 3-8 Cutaway diagram of a package ion exchange system. SOURCE: The Graver Company, Graver Water Division Union, New Jersey.

sodium. At pH less than 9.3, which is typical of treated water, ammonia is present in an ionic form (NH_4^+) that can also be removed by cation exchange, thus reducing the possibility of microbial growth in the water distribution system. Cation exchange resins have also proven effective in removing barium (Ba^{2+}), radium (Ra^{2+}), cadmium (Cd^{2+}), lead (Pb^{2+}), and trivalent chromium (Cr^{3+}). In many cases, cation exchange is the method of choice for radionuclide removal. It should be noted that adding sodium to water may not be desirable because of the need of some consumers to restrict their sodium intake.

Appropriate Water Quality and Performance Capabilities

Water to be treated by ion exchange must be low in solids to avoid fouling the resin. In particular, reduced iron species in ground water may become oxidized when the water is exposed to oxygen in the atmosphere and form precipitates that can damage the resin.

A resin may preferentially remove certain ions from solution. In general, it will remove highly charged ions more easily than it will monovalent ions. Calcium, magnesium, and reduced iron ions will be removed preferentially to other cations in a cation exchange system. If the target ion is other than these, the presence of such species may reduce removal efficiency.

Monitoring and Operating Requirements

An operator must monitor the system to determine the extent of resin saturation or the breakthrough of the ion to be removed. On ion breakthrough, the resin must be removed from service and regenerated. Ion exchange units can be controlled automatically, freeing the operator to make daily visits (rather than attending to the systems full time) to assure proper operation. However, determination of regeneration timing and troubleshooting requires an operator at the intermediate level of experience.

In either the cation or anion exchange process, the regeneration solution, which contains high concentrations of the undesired ions, must be carefully disposed of. Disposal can be quite costly, especially in the case of concentrated radionuclides.

Suitability for Small Systems

Ion exchange processes can be used with fluctuating plant flow rates and are readily adaptable to small units. Ion exchange is a common water treatment technology, available in point-of-use and point-of-entry devices as well as full-scale treatment plants. It is readily adaptable to small treatment plants.

An example of a small system that uses an ion exchange system is one serving approximately 400 people in Blue Mountain Subdivision, near Denver, Colorado. This system uses ion exchange for removal of uranium from ground water. Uranium levels in the raw water have been as high as 135 pCi/liter. The finished water levels are typically less than 1.5 pCi/liter, well below the drinking water standard for uranium of 30 pCi/liter (Tamburini and Habenicht, 1992). Similar beneficial results have been obtained using ion exchange for radium-226 removal from ground water near Spicewood, Texas, at the Quail Creek water system serving approximately 200 people (McKelvey et al., 1993).

Activated Alumina

How the Process Works

Activated alumina is useful for removing negatively charged ions. Activated alumina displays amphoteric properties, meaning its surface charge changes with solution pH. Alumina is not charged at a pH of 9.5, is positively charged below this pH, and is negatively charged above it. When treated with an acid solution, alumina is strongly positively charged and will select highly for fluoride (F^-), selenium species (SeO_4^{2-}, $HSeO_3^{2-}$), and arsenic($H_2AsO_4^-$). The greatest adsorption capacity for fluoride occurs at pH 5.5.

Appropriate Water Quality and Performance Characteristics

Water quality strongly influences the residence times and flow rates necessary for proper operation of activated alumina columns. In particular, the potential for preferential exchange of anions other than the target compounds in the raw water must be evaluated for each source water that will be treated with activated alumina. Therefore, pilot studies are an essential part of the design and evaluation of a full-scale activated alumina exchange system.

Monitoring and Operating Requirements

Regeneration of alumina requires a sodium hydroxide (NaOH) solution to remove the anions from the mineral surface. Following regeneration, the alumina column must be returned to the acidic state by rinsing with raw water followed by an acid solution. Anions to be removed from solution must compete with other anions, such as sulfate and hydroxide, for adsorption sites on the alumina. For this reason, sulfuric acid should not be used as the acidifying solution. In addition, the alumina dissolves slightly in sodium hydroxide. Over time the media will dissolve and require replacement.

As with ion exchange, regenerant disposal can be a problem. Some facilities discharge the regenerant brine solution to lined evaporation ponds designed for this purpose. After the water evaporates from the salts, the dried salts are disposed of in a landfill. Disposal costs can make up much of the operating cost of this technology.

Generally, the cost of an activated alumina system, including capital and operating costs, is quite high compared to other water treatment processes. In addition, operation of these systems requires advanced knowledge of water treatment principles and practice. Few systems are in operation at full scale, possibly due to cost and operating factors.

Suitability for Small Systems

Despite its cost and operational complexity, activated alumina can adapt readily to a small system. Columns can be scaled to fit any influent flow rate. Expansion can be accommodated by adding additional columns.

Activated alumina is in use in several plants in the southwestern United States. For example, the system in Desert Center, California, serves 10,000 people and uses activated alumina to lower the fluoride level to less than 1 mg/

liter from 8 mg/liter in the source water. The plant at X-9 Ranch near Tucson, Arizona, delivers water with significantly lowered fluoride levels to its 4,500 customers (Sorg, 1978). Both plants have decades of experience with the technology.

TECHNOLOGIES FOR SYSTEMS WITH
SURFACE WATER SOURCES

Historically, the primary concerns when treating surface water have been inactivation of microbial contaminants to prevent the spread of waterborne disease and removal of turbidity to make the water more palatable and to ensure that particles that may harbor microorganisms are not conveyed to the consumer's tap. As a consequence, the EPA's Surface Water Treatment Rule (SWTR) requires disinfection of all surface waters and filtration for most surface waters before distribution to customers. The total extent of inactivation and physical removal must equal 3 logs (99.9 percent) for *Giardia* cysts and 4 logs (99.99 percent) for viruses. In the future, it is likely that new regulations will also require *Cryptosporidium* removal.

Under the SWTR, the EPA gives water utilities "log credits" for inactivating pathogens when they use certain standard treatment processes, so that the utilities need not monitor their water for the viruses and parasites themselves. The supplementary information published with the SWTR specifies log credits to be given for both chemical disinfectants and physical filtration processes. The four filtration processes referenced in the SWTR are (1) conventional filtration, (2) direct filtration, (3) diatomaceous earth filtration, and (4) slow sand filtration. The SWTR defines conventional filtration and direct filtration to include chemical coagulation, flocculation, and, in the case of conventional filtration, sedimentation ahead of the filtration process. Several variations on treatment involving coagulation and filtration can be found in preengineered or package plants. State drinking water regulatory agencies would be responsible for deciding whether the package plants qualified for the log removal credits allotted to conventional filtration or to the reduced credit allotted to direct filtration. Filtration processes that do not function on the principles of the four defined processes listed in the SWTR are called alternative filtration processes. Examples include bag filters and cartridge filters. The logs of removal for *Giardia* cysts or viruses that can be allowed for alternative processes must be determined for each process. Requirements for demonstrating microbial removal vary from state to state because application of alternative filtration technology is subject to approval by the individual states.

Table 3-3 outlines the contaminants treated by various surface water filtration processes (discussed in detail below), the state of the technology, and a relative estimate of capital costs. Table 3-4 lists operating characteristics. The tables and the following discussion group conventional and direct filtration under the heading "coagulation/filtration."

Coagulation/Filtration

How the Process Works

Coagulation/filtration processes employ chemicals such as iron salts, aluminum salts, or cationic polymers to coagulate and destabilize suspended solids in the water so they can be removed by sedimentation and filtration. The processes defined in the SWTR use some form of flocculation, meaning slow agitation of the water to promote formation of larger flocs following the addition of coagulants. A solids removal step such as sedimentation may be used. The particulate matter remaining in the water is then removed by deep bed filtration. The filtration step works because the dose of coagulant chemical destabilizes the small particles in the water so that they attach to grains of filter material in the deep bed. Most of the particulate matter removed in coagulation/filtration is trapped in the filter bed by surface attachment mechanisms; only a small portion is strained or screened. Deep-bed, rapid-rate filtration without the use of coagulant chemicals does not qualify as a defined filtration process under the SWTR and thus is an alternative filtration process. Coagulation/filtration treatment plants are not effective if the coagulation chemistry is incorrect. When coagulation is correct, however, these plants are effective and versatile.

Appropriate Water Quality and Performance Capabilities

The range of water quality that can be treated by coagulation and filtration depends on the process train, specifically on the extent of solids removal provided ahead of filtration. Direct filtration has the most restricted range of water quality for which it can be applied because all solids must be removed in the filter bed. As the amount of particulate matter in the flocculated water increases, run length decreases.

Some suggested guidelines can be given for direct filtration. In the hands of a small system operator, direct filtration is not appropriate for treatment of water in which the average turbidity exceeds 10 NTU or the maximum turbidity exceeds approximately 20 NTU. Source water quality should be relatively stable. If raw water turbidity can increase by a factor of 10 in one day's time, the direct filtration process may not be appropriate. Two other important raw water quality characteristics are color and algae. Color removal requires doses of coagulant chemical related to the amount of color present. Thus an upper limit of approximately 40 color units would be appropriate for direct filtration (AWWA Committee, 1980). Because there are many species of algae and their effect on filtration differs according to species, no set numbers can be given for algae concentrations that could be treated by direct filtration. Algae removal must be evaluated on a case-by-case basis. Detention times in direct filtration plants are short, so added storage may be needed for disinfectant contact time.

Removing solids prior to filtration extends the range of water quality that can be treated by coagulation and filtration. Solids removal steps that increase particulate removal but are not as versatile as sedimentation include the various upflow and downflow flocculation/filtration or "roughing filter" processes. These processes employ some type of coarse medium in which flocculation occurs as a result of the mixing caused by the twists and turns the coagulated water must make as it passes through the bed. In addition, some solids removal occurs in the coarse medium, reducing some of the load on the filter. The greatest solids removal capability in pretreatment is attained by using sedimentation, which in small package plants is usually in the form of tube settlers. These plants may be able to successfully treat water with a turbidity of 200 NTU or perhaps higher or a color of 100 to 200 color units. As with other coagulation/filtration processes, the extent of algae removal is likely to be site specific and will depend on the type and concentration of algae present.

Coagulation/filtration has proven capable of removing turbidity, color, disinfection byproduct precursors, viruses, bacteria, and protozoa such as *Giardia* cysts and *Cryptosporidium* oocysts. Well-operated coagulation/filtration processes can produce filtered water with a turbidity of 0.10 NTU. Color removal depends on the pH during treatment and the coagulant dose employed. Removal of bacteria and protozoa can be as high as 3 to 4 logs (99.9 to 99.99 percent). Viruses are more difficult to remove, but Robeck et al. (1962) demonstrated 1-log (90 percent) to 2-log (90 to 99 percent) removal of poliovirus for direct filtration and removals exceeding 2 logs for conventional treatment.

Among the most challenging conditions for treatment by coagulation/filtration are very cold water, approximately 5°C or colder, and turbidities of approximately 10 NTU and lower. When the amount of particulate matter in the water is low, sedimentation is not very effective. Another difficult condition for water treatment is the combination of high color and moderate to high turbidity. The pH that is best for color removal may be different from the pH that is best for turbidity removal. In such a situation, identifying chemical conditions for optimum coagulation and filtration may be difficult. Finally, as mentioned previously, presence of algae in the raw water can make treatment difficult because some algae clog filters and cause very short filter runs.

Monitoring and Operating Requirements

Monitoring requirements include turbidity and pH measurement. The SWTR requires that filtered water turbidity be monitored every 4 hours, although this may be reduced to once per day for systems serving 500 or fewer persons, with state regulatory agency approval. Both streams and small reservoirs are subject to rapid changes in water quality, especially as a result of heavy rainfall in the watershed. Therefore, turbidity monitoring frequencies of once per 4 hours or once per 24 hours are minimal and are certainly not sufficient for effective operation of coagulation and filtration when water quality is changing rapidly. Continuous monitoring of filtered water turbidity is much better because the output from the turbidimeter can be used as an aid to controlling the plant if the operator is absent. In addition to monitoring of the filtered water, raw water turbidity and the turbidity of any solids removal process ahead of filtration should be checked at periodic intervals, such as every 4 hours, when the plant is operating. Raw and treated water pH should be monitored at least once per day because of the importance of coagulation pH for turbidity and color removal.

Process equipment should be designed so that flows of raw and treated water, filter head loss, and chemical feeds can be monitored easily by the operator. In addition, sample taps should be provided so the operator can obtain samples of raw water, pretreated water, and filtered water for analysis.

When equipment is entrusted to part-time operators, the foremost operating requirement for any small system treatment process is simplicity and ease of operation. Although coagulation/filtration processes have many excellent treatment capabilities, the complexity of coagulation chemistry does not decrease with the size of the treatment plant. Therefore, small systems can face great difficulties in managing coagulation because their resources in terms of operator training and experience are in most cases limited. Equipment manufacturers have attempted to help small system operators overcome the difficulties of coagulation by providing instrumentation that can be used to control some aspect of plant operation. For example, some package plants use continuous turbidimeters to adjust the coagulant chemical dose upward if raw water turbidity rises to a predetermined level. Some package plants have continuous turbidimeters on the effluent, and these cause the filter to be backwashed if treated water turbidity exceeds a preset value. Other package plants use a streaming current detector to adjust the coagulant feed pump.

A second critical need for coagulation/filtration systems is continuous operation at uniform flow rates. This is the ideal mode of operation for deep-bed granular media filters. Increases in filtration rate or start-stop operation can force previously trapped floc through the filter bed and into treated water. Discharge of floc in this manner can also cause the discharge of pathogenic organisms into the treated water, with the attendant increased risk of waterborne disease. After a deep-bed filter has been shut off, it should be backwashed to clean out the floc trapped in the bed. If this is not done, floc may be discharged into the treated water when the filter is restarted. Changes in filtration rate occur at most filtration plants, but start-stop operation is probably much more common in small systems, some of which try to produce enough filtered water in a single shift to last for an entire day.

Suitability for Small Systems

Numerous variations of the coagulation/filtration treatment train are presently being marketed as package plants (see Figure 3-9). A key factor in making affordable coagulation/filtration systems is the use of high-rate sedimentation or solids removal processes and use of filtration rates on the order of 12 to 17 m/h (5 to 7 gpm/sq ft).

One approach to coagulation and filtration in package plants involves chemical addition and optional in-line mixing, followed by flocculation, sedimentation in tube settlers, and multi-media filtration. The detention time in this process train is approximately 50 minutes and is longer than the detention time in some other package plants involving coagulation and filtration.

Another approach used by a number of manufacturers involves chemical addition and optional in-line mixing followed by a "roughing filter" (a unit process given different names by different manufacturers), followed by a multimedia filter. Detention times in these units may be on the order of 10 to 20 minutes, which definitely would not be sufficient for disinfection contact time when free chlorine is the disinfectant. Because of the relatively short detention times in package plants, many small systems treating surface water with package plants may need to provide for separate treated water storage facilities at the plant site to attain adequate *CT* values.

The availability of package plants has encouraged application of coagulation/filtration technology to small systems. Installation and operation of a coagulation/filtration package plant was described by Brigano et al. (1994). A key aspect of this application was use of a telemetry device to relay operating data to an off-site office of a contract operator. This substantially lowered the number of hours an operator needed to spend at the treatment plant.

Coagulation/filtration process trains can cope with a wider range of surface water quality than other filtration process trains, so they would seem logical choices for treating many surface waters. However, coagulation/filtration technology requires careful monitoring and oversight, whether by an operator at the plant or by remote sensing and data transmission from another location. Poorly operated coagulation/filtration technology of any size can be ineffective for treating water. Poor operation of such plants has resulted in numerous waterborne disease outbreaks. Small water systems that employ this technology must make a commitment to sustained excellence of operation.

Dissolved Air Flotation

How the Process Works

Dissolved air flotation (DAF) is most useful for removing particulate matter

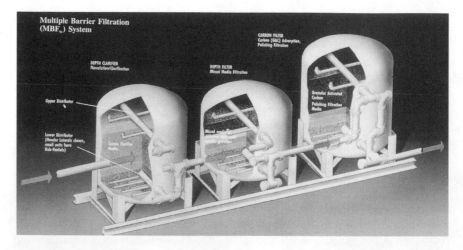

FIGURE 3-9 Example of a package filtration system. The first unit shown is a "roughing filter," used for flocculation and removal of some solids. The second unit is a multimedia filter, which contains layers of granular material to filter out particles remaining after the roughing filter. The last unit is a GAC filter (an optional add-on to this system) for removing dissolved organic compounds. SOURCE: Culligan International Company, Northbrook, Illinois.

and flocculated material that do not readily settle. The technology is a variation of coagulation/filtration, and therefore much of what applies to conventional coagulation/filtration systems also applies to DAF.

In the DAF process, raw water is coagulated and flocculated. Flocculated water flows to a basin where the floc is floated to the water surface by a cloud of microscopic bubbles, in contrast to conventional treatment employing a sedimentation process in which the solids settle to the bottom. The solids separation step in pretreatment with DAF is as effective as the solids separation step in pretreatment with sedimentation (conventional treatment), within appropriate source water quality limits.

The flotation action in DAF is caused by injecting water containing air dissolved at high pressure into flocculated water as it enters the flotation basin. This water, called the recycle stream, constitutes approximately 5 to 10 percent of the process flow. Recycle water is withdrawn from the bottom of the flotation basin, pumped into a pressure vessel (saturator) at 350 to 500 kPa (50 to 70 psi), and then returned to the flotation basin through a valve that dissipates the pressure. After the pressure returns to atmospheric pressure, the air dissolved in the recycle water comes out of solution in the form of microscopic bubbles. The air bubbles grow and rise to the top of the flotation basin, carrying the floc up to the surface where it can be skimmed off. Thus, the DAF process is an alternative to sedimentation.

Appropriate Water Quality and Performance Capabilities

The use of DAF as a pretreatment step before filtration has advantages over gravity sedimentation for treating algae-laden waters, highly colored waters, waters with low turbidity and low alkalinity, waters supersaturated with air, cold waters, and waters in intermittently operated treatment plants (Kollajtis, 1991).

DAF is best suited to removal of floc having low density because the floc must be floated to the surface. An example of this is floc formed by coagulation of color in low-turbidity water. DAF has been used to treat the algae-laden effluent from wastewater stabilization ponds, so it is not likely that an upper limit would apply on algae concentrations for potable water treatment.

When clay and silt are present in the source water, floc formed by coagulation is denser and not easily floated. Therefore, DAF is not an appropriate technology for treatment of turbid raw waters. An upper limit for turbidity might be in the range of 30 to 50 NTU for small systems, although Kollajtis (1991) has suggested that DAF may be applicable to waters having turbidity up to 100 NTU.

Malley and Edzwald (1991) compared DAF to conventional gravity sedimentation and found that for treatment of low-turbidity source waters, DAF was superior for removal of turbidity, and its performance for removal of total organic carbon, true color, and dissolved organic halide precursor materials equaled that of sedimentation. This reinforces the concept that a water treatment plant employing DAF for solids separation in pretreatment should be considered the equivalent of a conventional treatment plant for regulatory compliance purposes. The very short detention times in flocculation and flotation, however, mean that storage may be needed after filtration to increase the disinfectant contact time.

Hall et al. (1994) evaluated DAF for removal of *Cryptosporidium* oocysts. Their studies suggest that a treatment train consisting of chemical coagulation, flocculation, DAF, and filtration should be capable of removing 3 logs (99.9 percent) of the oocysts.

Monitoring and Operating Requirements

Monitoring needs for DAF are similar to those for conventional coagulation/ filtration systems. Raw and filtered water turbidity and pH and filter head loss should be monitored at the same frequencies as those employed in conventional treatment plants. Process equipment should be designed so that the operator can easily monitor the flow of raw and treated water, head loss, and chemical feeds. In addition, sample taps should be provided so the operator can obtain samples of raw water, pretreated water, and filtered water for analysis.

Additional monitoring, beyond that required for conventional coagulation/ filtration processes, is needed for DAF to control the air dissolution step. The key factors are saturator pressure and flow rate for the recycle stream. These must be metered to enable the operator to control the recycle step. In addition, the plant operator should periodically observe the condition of the floc that has floated to the surface of the flotation basin, as well as the nature of the bubbles formed in the basin. Vigorous, turbulent bubbling action is a sign of problems with the DAF process, and this must be avoided because excessive turbulence can break up floated floc and cause it to sink into the flotation basin, from which it could be discharged to the filters.

Suitability for Small Systems

Preengineered package treatment plants using DAF are available and have been used for more then a decade in the United States and even longer in Europe. To provide for economical and affordable small treatment plants, DAF package plants commonly combine flotation and filtration into one process basin. This is feasible because the solids separation step carries the floc to the surface of the water, producing clarified water at the bottom of the basin, which can be used to provide space for filter media and underdrain facilities. Two treatment steps are accomplished in the same space, resulting in substantial economies. However, combining the two processes results in operation of the DAF process at an over-flow rate that is the same as the filtration rate for the plant. This tends to place an upper limit of 5 to 10 m/h (2 to 4 gpm/sq ft) on the DAF overflow rates. In addition, filter backwashing would interrupt the DAF process, but because DAF can produce good treated water very quickly upon start-up, this may not be a problem. Thorough flocculation is essential in these systems because it is not possible to improve filtration by adding a filter aid.

The DAF process is more complex to operate than a conventional coagulation/filtration system because of the need to control the recycle flow stream and saturator operation. Failure of either the air saturation or flow recycle steps will cause the flotation step to fail. If this happens, all of the coagulated and floccu-lated solids have to be removed by the filter, in a process analogous to direct filtration. However, DAF is the best process for treating raw water with high concentrations of algae, and it is excellent for treating soft, highly colored waters with low-turbidity waters. For these reasons, it may find application in some small systems, despite its complexity.

Diatomaceous Earth Filtration

How the Process Works

Diatomaceous earth (DE) filtration is used primarily for particulate contami-nant removal. Industries have long used the process for filtration of liquids. The technology was developed for potable water treatment during World War II. Because of the need for portable water treatment equipment, the U.S. Army developed DE filters that could be mounted on trucks and transported to field locations. The portable size of these units makes them appropriate for small systems.

DE filtration works by straining particulate matter from the water. Coagulant chemicals are rarely used. Filtration is accomplished at the surface of a cake of diatomaceous earth (a fine-grade material composed of the fossil remains of diatoms) placed on filter leaves, or septa. This cake, called precoat, is established on the filter by recirculating a slurry of DE through the filter. After the precoat forms on the filter leaves, raw water containing some diatomaceous earth (body feed) is fed through the filter.

During a filter run, removal of particulate matter in the raw water, plus the accumulation of body feed diatomaceous earth material, causes the head loss to build up in the filter. When terminal head loss is reached, the flow of water into the filter is stopped and the filter is cleaned. The diatomaceous earth removed from the filter leaves is discarded.

There are two types of DE filters: (1) pressure filters, which have a pump or high- pressure water source on the influent side, and (2) vacuum filters, which have a pump on the effluent side. Vacuum filters are open to the atmosphere. Pressure filters are enclosed within pressure vessels.

Appropriate Water Quality and Performance Capabilities

Raw water quality should be excellent, with an upper limit of approximately 10 NTU (Letterman and Logsdon, 1976). Because DE filtration usually does not involve coagulation, capability for removal of dissolved constituents, such as color and inorganic contaminants, is very low. Thus, it is very important to determine in advance the quality of the raw water to be treated by DE filtration.

The size of the particles removed by DE filtration is a function of the size distribution of the diatomaceous earth particles used for the precoat and body feed. Fine grades of diatomaceous earth can remove smaller particles, such as bacteria. The grades of diatomaceous earth commonly used in potable water treatment are very effective for removal of *Giardia* cysts and *Cryptosporidium* oocysts. Schuler et al. (1988) reported removals exceeding 4 logs (99.99 percent) for both *Giardia* and *Cryptosporidium*. DE filtration is not as effective for bacteria removal, and it does not remove viruses very well unless the diatomaceous earth has been specially treated to alter its surface charge and bring about attachment of viruses to the diatomaceous earth. DE filters can remove algae to a very high degree, but the accumulation of algae cells on the surface of the filter cake can cause rapid clogging, so care must be taken to avoid excessive algae when applying DE filtration. Syrotynski and Stone (1975) reported that DE filters subjected to raw water containing microscopic total counts of 3,000 areal standard units per milliliter would experience shorter filter runs. Because DE filters have short detention times, disinfection contact time is necessary after filtration.

Monitoring and Operating Requirements

Monitoring requirements for DE filtration are simpler than those for coagulation and filtration because coagulant chemicals are hardly ever used. Raw and filtered water turbidity should be monitored, with compliance monitoring for filtered water turbidity done every 4 hours except for systems serving 500 or fewer people, which, after obtaining state approval, may monitor only once per day. If the nature of the turbidity-causing particulate matter remains stable, it may be possible to establish a ratio between the raw water turbidity and the appropriate dose of diatomaceous earth for use in the body feed. This situation might apply to treatment of lake water, for example. Filter head loss monitoring is necessary so the operator can determine when to backwash the filter. In addition, monitoring head loss can help establish the appropriate body feed. If water quality suddenly changes and the rate of head loss accumulation increases, more body feed may be needed. In addition to monitoring the flow through the DE filter, the operator needs to monitor the flow rate for body feed addition in order to control this aspect of plant operation.

In general, DE filter plant operators need mechanical skills to operate the body feed pumps, precoat pumps, mixers, and pipes and valves. Keeping the filter leaves clean in a DE filter is of primary importance. A filter leaf that is not properly cleaned at the end of a filter run can accumulate dirt and slime on the filter cloth, and this can prevent the formation of a uniform precoat when the filter is restored to service. DE filtration equipment should be designed so that the plant operator can easily inspect the cleanliness and integrity of the filter leaves.

One problematic aspect of small DE filter plants is the tendency of many small systems to operate filters intermittently rather than on a 24-hour-per-day basis. Unless provision to continuously recirculate filtered water through the DE filter is provided, every time the filter is stopped, the filter leaves must be cleaned and the used diatomaceous earth thrown away. When DE filters are capable of having continuous runs lasting as long as 2 to 4 days, wasting the precoat and body feed diatomaceous earth at the end of a run as short as 8 hours can drive up operating costs. Used or spent diatomaceous earth must always be cleaned out of the filter, or contaminants trapped in the filter cake may pass through the filter and into the treated water in a subsequent filter run.

Suitability for Small Systems

DE filtration is well suited to small systems and has been used in the past by such systems. In a survey of direct filtration, Letterman and Logsdon (1976) reported that among the 13 DE filter plants responding to the survey, 4 served approximately 3,300 to 4,800 people, and another 4 served approximately 6,700 to 20,000 people. A key factor in the use of DE filtration for small systems is that chemical coagulation is not necessary, so operators do not have to learn about this complex aspect of water treatment. Waters suitable for DE filtration are low in turbidity and in color or other organic matter that can form disinfection byproducts when chlorinated.

Slow Sand Filtration

How the Process Works

In slow sand filtration, biological action breaks down some organic matter, and some inert suspended particles are physically removed from the water. Slow sand filtration was the original form of water treatment used by municipalities in the nineteenth century and is now considered a low-technology approach to water treatment.

The SWTR defines slow sand filtration as ". . . a process involving passage of raw water through a bed of sand at low velocity (generally less than 0.4 m/h) resulting in substantial particulate removal by physical and biological means" (EPA, 1989). In this process, uncoagulated water is applied to a bed of sand having an effective size of approximately 0.3 mm and a depth of approximately 0.6 to 1.2 m (2 to 4 ft) at a filtration rate of 0.1 to 0.4 m/h (0.04 to 0.16 gpm/sq ft). With extended use of the filter, a biological ecosystem grows in the sand bed. On the top of the filter media, a biologically active organic layer (known by the German term *Schmutzdecke*) builds up and assists filtration. The water then enters the top layer of sand, where more biological action occurs and particles attach to sand grain surfaces by adsorption and sedimentation in pores between sand grains.

The biological activity within the sand bed is a key factor in the effective action of slow sand filters. Biota in slow sand filters include bacteria, algae, rhizopods, ciliates, rotifers, copepods, and aquatic worms (Haarhoff and Cleasby, 1991). Fresh, clean sand is not as effective as a "ripened" sand bed that has been in service long enough for the ecosystem to become established. Depending on the available nutrients in the source water and the water temperature, establishing the ecosystem could take from a few weeks to 2 to 3 months. Operation and maintenance activities that harm or inactivate the ecosystem therefore tend to cause slow sand filter performance to deteriorate.

Providing for storage of filtered water is essential at a slow sand filter plant for two reasons. First, because of the importance of establishing biological activity, using chlorine ahead of the filter is inappropriate, and disinfectant contact time must be provided in a storage basin after filtration. Second, storage is needed for equalization of production and demand. Slow sand filters should be operated at steady rates, if possible, and flows should not be increased or decreased frequently to keep pace with system demand. In very small systems, the need for disinfectant contact time plus equalization storage could require the provision for storage of approximately 1 day's production at the plant.

Appropriate Water Quality and Performance Capabilities

Because slow sand filters in the United States generally are used without pretreatment, the range of raw water quality appropriate for treatment by this process is rather narrow. Cleasby (1991) recommends the following guidelines for ideal source water quality for slow sand filtration without pretreatment:

- turbidity   < 5 NTU
- algae   no heavy seasonal blooms; chlorophyll-a < 5 µg/liter
- iron   < 0.3 mg/liter
- manganese   < 0.05 mg/liter

Source waters with clay content may cause treatment problems. Fox et al. (1984) operated a slow sand filter to treat Ohio River water (a clay-bearing water source) and found that although the influent turbidity ranged from approximately 10 to 23 NTU for the first 50 days and then was between 10 and 0.4 NTU for the next 130 days, filtered water turbidity was progressively poorer during the filter operation and eventually exceeded 1 NTU. In addition, the length of each filter run became shorter, from 98 days for the first run to 6 days for the last, indicating progressive clogging of the sand bed with clay .

In most waters, slow sand filters can reduce turbidity sufficiently to satisfy regulatory requirements, but in others, turbidity reduction may be minimal. Turbidity removal may be impaired in waters with very low nutrient content (Bellamy et al., 1985b), as some nutrients must be present to promote growth of the biological ecosystem within the filter bed.

Algae in raw water can clog slow sand filters. Cleasby et al. (1984) found that when chlorophyll-a (an indirect measure of algal concentration) was between 8 and 138 µg/liter, four filter runs varied in length between 10 and 22 days. Filter runs were 34 to 123 days when chlorophyll-a was in the range of 1 to 4 µg/liter.

Slow sand filters are not very effective at removing disinfection byproduct precursors or color. If the biological action within a filter bed were effective for removal of organic matter of this type, biological action in lakes and rivers would have already removed the organic matter from the source water.

Slow sand filtration excels at removing microorganisms. Its effectiveness for this purpose was demonstrated in the nineteenth century by the reduction in waterborne disease in European and English cities that used slow sand filtration. Research in the twentieth century has documented the efficacy of slow sand filtration for virus removal (Poynter and Slade, 1977), *Giardia* cyst removal (Bellamy et al., 1985a,b; Pyper, 1985; Seelaus et al., 1986; Schuler et al. 1988), and *Cryptosporidium* oocyst removal (Schuler et al., 1988). Slow sand filters are less effective at removing microorganisms from cold waters because as temperatures decrease, the biological activity within the filter bed declines. For this reason, slow sand filters that will treat water at temperatures below approximately 10°C should be conservatively designed; i.e., filtration rates should be near 0.12 to 0.17 m/h during winter operation (Pyper and Logsdon, 1991).

One modification of slow sand filtration that offers promise for removal of organics is the GAC sandwich filter. This filter uses a base sand layer of approximately 30 cm, an intermediate GAC layer of approximately 15 cm, and a top sand layer of approximately 45 cm. This modified slow sand filter has been effective for removal of pesticides, total organic carbon, and trihalomethane precursors (Bauer et al., 1996).

Monitoring and Operating Requirements

Monitoring and operation of slow sand filters is not complicated. Daily tasks include reading and recording head loss, raw and filtered water turbidity, flow rates, and disinfectant residual. If necessary, flow should be adjusted to bring water production in line with demand. In addition, with the promulgation of the SWTR, each day the operator would need to use the flow data and disinfectant residual data to calculate *CT* values and determine if disinfection is sufficiently rigorous. These duties may require 1 to 2 hours unless automated.

As head loss increases in the slow sand filter bed, eventually the filter will need to be cleaned. This is accomplished by draining the filter and removing 1.2 to 2.5 cm (0.5 to 1 in.) of sand from the top of the bed. In a study of slow sand filter operation and maintenance, Cullen and Letterman (1985) estimated that approximately 5 hours would be required to scrape 100 m^2 of sand bed. After repeated scrapings, so much sand will have been removed that replacement of sand is necessary. This replacement, known as "resanding", is labor intensive. Cullen and Letterman (1985) estimated that resanding a depth of 15 to 30 cm would require 48 to 59 hours of labor per 100 m^2 of filter bed. These values would be modified somewhat if more machinery were used.

Suitability for Small Systems

Slow sand filtration has been adapted to package plant construction. Hall and Hyde (1987) reported on a project to evaluate a slow sand filtration package plant consisting of two separate 2-m^2 filters, each with a raw water inlet, two flow controllers, a chlorine feeder, a chlorine contact tank, and a service reservoir. The package slow sand filter produced filtered water turbidity averaging less than 1 formazin turbidity unit (FTU) through the study, and filtered water turbidity remained at less than 2 FTU even when raw water turbidity was as high as 94

FTU. Average coliform removal by filtration was 94 percent. Fewer than 2 days per year were required for sand cleaning.

In a recent application of slow sand filtration technology, one small water system used precast concrete boxes as filter cells for a 300-liter/min plant (Riesenberg et al., 1995). The precast filter boxes could be tested for water tightness and repaired if needed at the manufacturing facility. Such an approach could provide both labor savings and improved quality control for construction of slow sand filter plants serving approximately 500 or fewer persons.

Slow sand filtration is among the simplest and most easily used of the technologies available for small water systems because the efficacy of the process is mainly dependent on the inherent mechanisms at work in the process rather than on the actions of the plant operator. However, because few remedies are available to a plant operator if slow sand filtration is ineffective, the process must be used with caution. Only high-quality surface waters (low in turbidity, algae, and color) are suitable for application to slow sand filters without pretreatment or process modifications such as the use of a GAC layer in the filter. When used with source water of appropriate quality, however, this process may be the most suitable choice for small systems that must filter surface water.

Bag and Cartridge Filters

How the Process Works

Bag filters and cartridge filters are technologies specifically developed for small to very small systems. They are made from pressure vessels containing a woven bag or a cartridge with a wound filament filter. Water passes through the bag or the wound filament cartridge, and the filter removes particulate matter large enough to be trapped in the pores of the bag or cartridge. The filters are appropriate for removal of *Giardia* cysts and possibly for removal of *Cryptosporidium* oocysts (which are large enough to be strained in the filter pores) but not for removal of bacteria and viruses. They are designed for simple operation; no coagulant chemicals are used.

Proper selection of the pore size of cartridge and bag filters is critical. Because cysts and oocysts are biological particles without hard shells or skeletons, they are capable of deforming somewhat and squeezing through pores that might seem to be small enough to prevent their passage. In addition, wound filament filter cartridges have pores that are both larger and smaller than the nominal size indicated in the equipment literature. Therefore, these filters do not provide an absolute cutoff for particles at or slightly larger than their nominal size.

Bag filters and cartridge filters function by surface straining, so a mat or cake builds up on the filter surface. If the materials being removed are not compressible, the buildup of this cake may not hinder filtration. Conversely, removal of compressible particles such as algae or fragments of biological matter can blind the filter. This same phenomenon can occur in DE filtration, which also involves a surface filtration mechanism. In some instances, decreasing the influent pressure on the filters can result in longer service life and greater throughput. This probably happens because the lower head loss through the filter causes less compression of the compressible particulate matter and thus reduces the tendency of compressible particles to blind the filter surface.

As water flows through a bag or cartridge filter, eventually the pressure drop within the filter builds up until it becomes necessary to terminate the filter run. When this happens the used bag or cartridge is thrown away and replaced with a clean one.

Appropriate Water Quality and Performance Capabilities

Because filtration of some types of particles can blind bag and cartridge filters, these filters are appropriate only for high-quality waters. In fact, source water quality for bag filters should be higher than the quality for slow sand filters. Source water turbidity may not be an adequate indicator of the water's suitability for treatment by bag and cartridge filters. Hard, mineral materials are not as likely to blind a filter as are biological particles such as algae and fragments of disintegrating biological matter. The number of gallons of water that can be filtered could vary by a factor of 10 or greater for water of a given turbidity, depending on the nature and concentration of particulate matter in the raw water.

Bag and cartridge filters merely strain particulate contaminants out of water, so they are not appropriate for removal of true color or other dissolved contaminants. Because they remove larger microbial contaminants such as protozoan cysts and oocysts, but are not particularly effective for removing bacteria and viruses, bag and cartridge filters are appropriate only for application to relatively pure surface waters, in which the concentration of bacteria and viruses that needs to be inactivated by disinfection is low. Bag and cartridge filters are not appropriate for treating source waters having elevated turbidity. They can remove approximately half of the turbidity in some raw waters, and in such cases if the raw water turbidity were greater than approximately 2 to 3 NTU, the filtered water turbidity would exceed 1 NTU.

Monitoring and Operating Requirements

Bag filters and cartridge filters are simple devices, and their monitoring requirements reflect this. Because of the SWTR requirements for turbidity monitoring, filtered water turbidity should be checked daily. Also, the operator should monitor head loss through the filter and total gallons of water filtered in order to estimate when the existing bag or cartridge will need replacement.

Operators should exercise care when changing filter bags or cartridges. The manufacturer's instructions on these procedures should be followed so that the clean cartridges or bags are not damaged on installation.

Disposal of bags or cartridges is simple because the filters do not remove toxic substances, so the spent bags or cartridges should be suitable for disposal to a landfill after they have dried. This is not expected to be a problem even if *Giardia* or *Cryptosporidium* is being removed, as the contents of a bag or cartridge filter should be no more microbiologically hazardous than the contents of a disposable diaper from an infant with giardiasis or cryptosporidiosis.

Because the requirements for virus removal and inactivation must be met entirely by disinfection in a treatment train involving a bag or cartridge filter, extra care is needed for this process. With the use of free chlorine, this is not difficult in most situations. The very short residence times in the filters, however, mean that disinfectant contact time would be needed in storage after filtration if sufficient contact time had not been attained before filtration.

Suitability for Small Systems

Bag filters and cartridge filters were developed specifically for small systems. The treatment capability of bag filters and cartridge filters is limited to removal of particles from water. They are capable of removing *Giardia* cysts and perhaps *Cryptosporidium* oocysts. Viruses not attached to other particulate matter would pass through these filters. In addition, these filters will not remove chemical contaminants present in solution. Bag filters and cartridge filters are most appropriate for treatment of very-high-quality source waters for removal of protozoan cysts and are best suited for very small systems, such as those serving fewer than 500 people.

CENTRALIZED OPERATION THROUGH AUTOMATION AND REMOTE MONITORING AND CONTROL

Some water treatment technologies respond well to automated operation. A major advantage of remote monitoring and control is the potential to share resources among several small systems, so that a single operator can monitor and operate several small plants in a given area. The operator can work from a centralized location and receive and respond to information from each plant. Perhaps more important, an operator with more training in water treatment can be employed because the group of small systems can share the higher salary requirements of an experienced operator. In addition, the automatic control of chemical feeders often lowers chemical costs and improves water quality.

Several levels of remote monitoring and control are available. The complexity will depend on the complexity of the treatment option and the availability of the operator. A basic monitoring system might include a simple auto-telephone dialer to alert the remote operator of such problems as power outages, pressure drops, unauthorized building intrusions, high sump levels (possible flooding), or any other condition that can be monitored by a simple on-off alarm. In their simplest form, such alarms are presented as a common alarm announcement, which then requires the operator to visit the site to determine the exact cause and nature of the alarm. Equipment operation is often performed by local hard-wired relay systems or individual control packages provided by the equipment supplier. There is no control system integration in this lowest level of monitoring and alarm.

A higher-level system would include an integrated approach, tying together the operation of the system and alarms. Different operational programs can address differing conditions. The operator can access the system remotely by computer modem to determine system status or to check the condition of an alarm; many systems also allow for the operator to remotely control system operation using the same computer modem.

The highest-level system of remote monitoring and control involves one (or more) master locations in constant communication with a number of remote unattended locations. The master location is often staffed on a full-time basis. This highest level of monitoring and control is sometimes referred to as supervisory control and automatic data acquisition (SCADA). The master operator can constantly monitor each remote system, adjusting operations at the master supervisory control console or dispatching personnel to the remote location as needed. A system of this sophistication can also allow data to be sent by telemetry to the central location for centralized performance of administrative tasks such as regulatory and management report preparation. Similarly, individual customer meter readings can be obtained for billing purposes.

Small systems may also operate without an attending operator in either of two automatic control options. In the simplest level of automatic control, sensors, instrumentation, and control devices operate on simple rules. An example is a chemical feed flow rate controller tied to the raw water influent flow rate. The feed controller sets a new chemical feed rate as the influent flow rate changes. Advanced automatic control relies on sophisticated computer models or artificial intelligence to make more advanced and/or precise corrections to system operation in response to changing water conditions. This type of advanced automated control is in its infancy.

The types of water treatment problems addressed at a given site will determine to a large extent the level of remote monitoring and control desired. If a short-term disfunction in a system could result in a high risk of an acute health effect such as breakthrough of *Giardia* or *Cryptosporidium*, or nitrate levels high enough to cause methemoglobinemia in infants, a high level of remote monitoring and control is advisable. A treatment system designed to protect the customer from a secondary, aesthetic water quality problem, such as colored water or excess iron or manganese, may need only a medium level of monitoring. In the extreme, a ground water well not influenced by surface water and located in a residential area may only need a light on the pumphouse to indicate pump failure.

The type of system management also affects the appropriate level of remote monitoring. A local operator responsible for a single location on a part-time basis may need only the simplest set of remote monitoring and alarm tools to respond effectively to system problems 24 hours a day. A centralized regional operation may require a higher level than normally needed at each remote location in order to reduce the number of required personnel. In either case, customer satisfaction is likely to be high. If the system is properly designed and operated, the operator will normally recognize and solve a problem before customers note a water quality deficiency.

A couple of caveats apply here. First, remote monitoring and/or remote operational control does not eliminate the need for maintenance. In fact, the increased reliance on sensors inherent in remote monitoring results in an increased need for sensor maintenance and calibration. Also, as these systems are developed, it is important that they conform to a standardized communication format. The electric industry has a utility communications architecture that is being integrated into an industry-government standard system for communication (Schlenger et al., 1994). Water industry control systems should begin to adopt these standards or develop another standard communication system in order to standardize data acquisition and reporting.

OPTIONS OTHER THAN CENTRALIZED TREATMENT

When a centralized treatment facility is not feasible and obtaining water from some other source is not possible, small systems may need to consider installing point-of-entry (POE) or point-of-use (POU) water treatment devices in their customers' homes or distributing bottled water. These alternatives generally are appropriate for system-wide use only for very small systems, particularly those serving 500 or fewer people.

Numerous households in the United States use POE and POU devices and bottled water, primarily to deal with aesthetic concerns. This report, however, discusses these options for purposes of providing water that meets the quality requirements of the SDWA. In such a situation, adoption of POE, POU, or bottled water as the means of providing drinking water is not an individual household's choice but the choice of the water system in cooperation with regulatory authorities. Therefore, circumstances surrounding use of POE, POU, or bottled water are much different than those related to the voluntary use of these options.

POE treatment devices are used to treat all water used in a household or building and result in water from any tap being suitable for drinking when treatment is effective. POU devices are used to treat the water at a single tap or faucet, and as a consequence, only that tap or faucet has potable water (see Figure 3-10).

If a POU device is placed under the kitchen sink to treat cold water for the kitchen faucet, only that water is potable; water from a faucet in a bathroom, a likely location for brushing teeth, would not be potable. This aspect of POU treatment has been a source of objection to its use.

POU and POE Treatment

Description

POE and POU systems often use the same technology concepts employed in centralized treatment, but the technology is applied at a much smaller scale and sometimes is modified for application to treatment of small flows. Most of the processes used in POE and POU units are discussed at length in previous sections of this chapter. The following are aspects of treatment technologies that are specific to application in POE and POU systems:

• *Activated alumina and granular activated carbon* typically are used in cartridges or pressure vessels for POE and POU treatment devices. Activated alumina treatment is most often used for fluoride removal. When the exchange/adsorption capacity of the activated alumina has been reached, the spent cartridge must be replaced. GAC systems are used for taste and odor concerns and for removal of regulated organic compounds. The performance and life of GAC systems depend on the amount of GAC used in the device, the contact time between the GAC and the water, and the contaminants being removed. When the GAC-treated water reaches a predetermined performance concentration for the contaminant being removed, the GAC must be replaced. This is done by removing the cartridge and installing a new one.

• *Reverse osmosis* devices for POE and POU need to be provided with a means of discharging reject water to a drain. The discharge line should be installed with an air gap so a cross-connection between wastewater and drinking water will not occur. Reverse osmosis and other membrane technologies are among the fastest-developing types of technology with possible applications for POE systems.

• When *ion exchange* technology is used in homes to soften potable drinking water, all of the household water generally is softened, and outside faucets used for lawn and garden watering might be unsoftened. Radium removal would be a possible POE application for this technology.

• *Air stripping* has been used in POE systems to remove volatile organic compounds and radon from ground water. For these applications, it is important to vent the off-gases adequately to avoid creating an air pollution hazard inside the home. Generally, this is achieved by designing the ventilation system such that the air duct for the vent disperses the stripped contaminants above the air envelope for the structure.

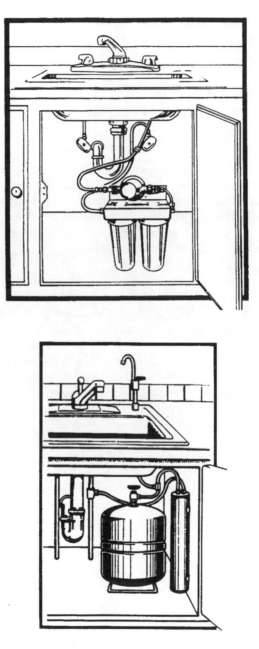

FIGURE 3-10 Examples of under-the-sink POU units. The top unit treats all the water flowing to the kitchen faucet. The bottom unit treats only water flowing to a special tap mounted beside the faucet. SOURCE: Reprinted, with permission, from Lykins et al. (1992). ©1992 by Lewis Publishers, Inc.

Because control of acute disease should be accomplished with the highest feasible degree of competence, use of POE and POU treatment for disinfection of surface water is not generally viewed as appropriate. In the future, however, disinfection may be required for ground water sources that currently are not disinfected. The safest, most effective, and most readily manageable disinfection method of POU/POE application for inactivation of bacteria and viruses is UV light. UV light as it is typically employed is not effective for *Giardia* cysts or *Cryptosporidium* oocysts, so it is applicable only to ground water. UV is currently the most popular disinfection method for POE and POU systems because it does not involve the addition of a chemical and therefore imparts no tastes, odors, or chemical byproducts (Lykins et al., 1992). UV systems range in capacity from 2 liters/min to approximately 2,000 liters/min; manufacturers claim effective life spans of 6,000 hours to 12,000 hours for the lamps used to produce the UV light.

Appropriate Water Quality

Although POE and POU systems may in some instances be used to treat surface waters, in a regulatory setting they would be appropriate only for ground water because of the frequency of monitoring (daily) necessary with surface water treatment and because of the necessity of ensuring thorough disinfection of surface water. The uniformity of ground water quality from a given well means less emphasis needs to be placed on monitoring because quality-related changes in treatment efficacy would not be as severe for ground water as they would be for most surface waters.

Selecting POE and POU equipment does not eliminate the need for evaluation of treatment efficacy before installing the treatment equipment. Before funds are expended to treat water for regulatory compliance purposes, verification of the efficacy of the proposed treatment technique is essential. For devices that employ cartridges (e.g., GAC columns or activated alumina), pilot testing of the source water may be necessary to develop valid estimates of the service life of the unit before replacement is required. Reverse osmosis testing would be done primarily to determine whether the water being treated will foul the membranes, as contaminant removal capabilities of a membrane do not vary from water to water. Ion exchange units for radium removal could be regenerated based on exhaustion of hardness removal capability, as radium is still removed after calcium and magnesium begin to appear in the product water. Before UV disinfection is used, testing for possible interferences to the transmission of UV light through the water would be advisable.

Monitoring and Operating Requirements

Effective operation, maintenance, and monitoring programs are essential to the overall performance of any water treatment system and are especially significant for POE and POU systems. Many homeowners assume their systems will perform properly once installed and do not understand the level of effort required to ensure proper operation. For this reason, when POE or POU systems are installed for regulatory purposes, programs for long-term operation, maintenance, and monitoring must be provided by water utilities or regulatory agencies.

Proper installation is the first step in effective long-term operation and maintenance of POE and POU systems. Installation must be done only by experienced contractors or installers whose products conform with applicable plumbing codes. Qualified installers carry liability insurance for property damage during installation, are accessible for service calls, accept responsibility for minor adjustments after installation, and give a valid estimate of the cost of installation.

After installation, POE and POU systems need a well-defined program of operation and maintenance for continued production of drinking water of acceptable quality. The equipment manufacturer's recommended operation and maintenance requirements can serve as the bases for the operation and maintenance program. Equipment dealers may provide maintenance for a limited time period as part of an installation warranty. A long-term maintenance program may be carried out by a local plumbing contractor, a POE or POU service representative or equipment dealer, a water service company, the local water utility, or a circuit rider (an individual under contract with several water systems to perform operation, maintenance, and monitoring activities) (Bellen et al., 1985). It is essential that maintenance be performed by personnel responsible to the small water system rather than to the homeowner because water system personnel will understand the need for a continuing operation and maintenance program, whereas some homeowners will not.

One way to ensure the production of water that meets regulatory requirements is to define a replacement schedule for media, cartridges, filters, and/or modules associated with POE and POU systems. Replacement schedules can be defined either by time (e.g., every 6 months) or by flow (e.g., every 30,000 liters). The advantage of using time is the avoidance of having to monitor flow. However, replacement based on time may result in equipment being replaced too early or too late. The former case would waste resources, while replacing equipment too late could result in the consumption of drinking water that exceeds one or more of the drinking water standards. Replacement based on flow requires that water meters be used as a part of the monitoring program. Although this approach requires a bit more hands-on involvement, it results in a better balance between maximizing equipment life and producing water that meets regulatory requirements.

Monitoring programs need to be site specific and reflect the contaminant or contaminants being removed, the equipment used, the number of POE or POU units in service, and the logistics of the service area. Minimum sampling frequencies and types of analyses should be established in cooperation with the local health department, the state regulatory agency, and the small system.

Monitoring programs generally include raw and treated water sample collection, meter reading, field analyses (measuring pH, dissolved oxygen concentration, and other parameters) as appropriate, shipment of samples to a laboratory, and recordkeeping. The use of state-approved sampling methods and certified laboratories is a requirement for regulatory compliance. Lykins et al. (1992) recommend that monitoring programs provide some way to respond to water quality questions from residents both with and without POE or POU systems and to assess raw water quality trends.

In addition to having samples collected by an employee of the small water system, options for sample collection include contracting with a POE or POU service representative, an independent laboratory, a local health department, a circuit-rider operator, or a trained community resident. An advantage of using a community resident or local representative is that these persons are familiar with the residents of the community and are likely to be better able to coordinate relatively convenient sample collection times. A disadvantage of using such a person is that community residents are likely to know the least about proper sample collection and preservation procedures, water quality tests, methods for recordkeeping, water meter reading, and proper procedures for transport or shipment of samples to an analytical laboratory. Training is necessary to enable a community resident to be an effective sample collector. Concepts related to training for sample collectors were presented by Bellen et al. (1985).

To avoid duplication of travel to homes and buildings equipped with POE or POU devices, the sample collector needs to be familiar with the treatment equipment used and the treatment objectives. An ability to conduct basic troubleshooting and to service equipment is also helpful, in case problems are brought to the attention of the sample collector during sampling rounds.

Monitoring of POE and POU treatment devices is problematic. When water is treated to meet MCLs or to satisfy treatment technique requirements, monitoring has to be done to verify that the water quality or treatment approach is satisfactory. From a regulatory agency perspective, monitoring of POE and POU devices is a major obstacle to acceptance. For a community consisting of 50 homes and served by a central treatment facility, regulatory compliance monitoring for most of the regulated contaminants could be done at the discharge point from the treatment plant or at the point of entry to the distribution system. If POE or POU devices were used instead of central treatment, the community of 50 homes would have 50 water treatment devices, any one of which might possibly malfunction or reach its capacity for effective treatment at some time. The oversight effort, both for the small water system and for the regulatory agency, is multiplied several fold in such a circumstance. The cost of monitoring every POE or POU device could be a burden on small water system customers.

One approach to lowering the cost of monitoring is to sample representative households that reflect typical POE or POU installations and levels of contamination rather than sampling all households with installed systems. The costs of monitoring would decrease as a smaller percentage of the devices was monitored in a year's time, but the risks of noncompliance with an MCL would increase. Striking a balance between the risks to persons consuming water exceeding MCLs because of insufficient monitoring and the cost of analyzing numerous monitoring samples will be a challenging task for small water systems using POU or POE devices and for the regulatory agencies overseeing such systems.

Regulatory Approach to POE and POU Systems

The EPA (1985) has established the following conditions that must be met to ensure protection of public health when POE or POU systems are used for compliance purposes:

• *Central control:* Regardless of who owns the POE or POU system, a public water system must be responsible for operating and maintaining it.
• *Effective monitoring:* A monitoring program must be developed and approved by the state regulatory agency before POE or POU systems are installed. Such a monitoring program must ensure that the systems provide health protection equivalent to that which would be provided by central water treatment meeting all primary and secondary standards. Also, information regarding total flow treated and the physical conditions of the equipment must be documented.
• *Effective technology:* The state must require adequate certification of performance and field testing as well as design review of each type of device used. Either the state or a third party acceptable to the state can conduct the certification program.
• *Microbiological safety:* To maintain the microbiological safety of water treated with POE or POU devices, the EPA suggests that control techniques such as backwashing, disinfection, and monitoring for microbial safety be implemented. The EPA considers this an important condition because disinfection is not normally provided after POE systems.
• *Consumer protection:* Every building connected to the public water system must install POE or POU treatment and adequately maintain and monitor it.

Although several states have developed regulations for the certification of POE and POU devices, California has the most extensive program for regulating the use of POE and POU systems in place of central treatment. The California action may be indicative of the approach other states will take in the future. The California Department of Health Services (DHS) does not allow the installation of POE or POU devices by community water systems unless all other available alternatives have been evaluated and found to be infeasible. The evaluation submitted to regulators must document the water quality problem or problems, alternatives pursued to correct the problem, potential for connection with an adjacent utility, comparison of POU or POE treatment versus central treatment, potential for development of new ground water sources, and potential for developing and treating a surface water source. In addition, the California DHS specifies a list of conditions that must be considered in the approval process for POE and POU devices. These conditions include utility responsibility for POE or POU ownership and maintenance, and for ongoing monitoring of contaminants, including monthly bacteriological samples. In addition, California regulations require that the POU and POE devices be either pilot tested at each individual site or that the performance of the equipment be certified in a formal testing program. Testing for certification must be conducted by a recognized testing organization and must be performed in an independent laboratory meeting laboratory accreditation requirements set forth by the California DHS. The testing must be carried out according to specified protocols accepted by the California DHS. If the equipment manufacturer makes health or safety claims regarding the ability of the device to remove specific contaminants, these claims must be verified. In addition, testing must demonstrate that the equipment will not add toxic substances to the treated water, such as by leaching from system components.

The California regulations for certification of POU and POE devices draw on standards for the testing of this equipment established by the National Sanitation Foundation (NSF) International. NSF International has issued seven standards related to the testing of POE and POU devices:

1. standard 42, which covers the ability of GAC and mechanical filtration to improve the aesthetic qualities of drinking water;
2. standard 44, which specifies testing protocols for cation exchange units;
3. standard 54, which provides protocols for testing the ability of GAC and mechanical filtration systems to remove contaminants posing a health hazard;
4. standard 55, which specifies how to test UV disinfection systems;
5. standard 58, which outlines testing requirements for reverse osmosis systems;
6. standard 61, which details how to test for the possibility that chemicals will leach from system components into the water; and
7. standard 62, which sets forth testing protocols for distillation systems.

NSF International has a certification laboratory that can conduct a full range of physical, microbiological, radiological, inorganic, and organic analyses.

The Water Quality Association (WQA) also has a certification program for POE and POU devices. However, the WQA is a trade association for POE and POU equipment manufacturers and therefore cannot provide the type of independent analysis available from NSF International (Lykins et al., 1992). Local planners considering the purchase of POE and POU devices need to be aware of this distinction when purchasing POE and POU equipment and interpret the WQA certification accordingly.

Circumstances for Use of POU or POE Systems

The drinking water industry and state regulatory agencies have often opposed the installation of POE or POU systems as the choice of technology to treat water and comply with drinking water regulations. Regulatory objections to these devices include the following:

• POU devices do not treat all the water taps in a house, posing the potential health risk of household residents drinking untreated water.

• Control of treatment, water quality monitoring, routine operation and maintenance, and regulatory oversight is complex because treatment is not centralized.

• Unless monitoring requirements are decreased from those stipulated for centralized treatment, monitoring is more costly than for centralized treatment because of the numerous individual home treatment devices that must be checked.

• Ensuring regulatory compliance is more difficult than with centralized treatment.

• Service life and efficiency of treatment units depend on source water quality, so performance can vary from household to household.

• Community water systems are concerned about the liability associated with entering a customer's home to monitor or service the units.

Despite these concerns, a driving force for the use of POU and POE treatment devices has been the cost differential. When POU devices are used, only water that is used for potable purposes is treated. If a source water is acceptable for drinking except for exceeding the standard for nitrate or fluoride, for example, treating the small number of liters per day needed for drinking and cooking might be less costly than installing a centralized treatment system that could remove nitrate or fluoride from all water used by the community. Water used to wash cars, water lawns, flush toilets, or launder clothing would not need to have nitrate or fluoride removed. Similarly, POE devices can save the cost of installing expensive new equipment in a central water treatment facility. They can also save the considerable costs of installing and maintaining water distribution mains when they are used in communities where homeowners have individual wells.

As the population served by a small system increases, the monitoring, operation, and maintenance costs associated with POU and POE devices increase in direct proportion to the population. Table 3-5 shows a cost comparison for using POE versus adding a GAC treatment system to the water treatment plant for a community with between 10 and 50 households (Goodrich et al., 1992). As the table shows, when 20 or more households are involved for this example, modifying the central treatment plant is less costly than installing and maintaining POE devices in individual homes. Figure 3-11 compares the cost of installing POE systems with that of connecting homes to a central water treatment plant. As

TABLE 3-5 Cost of POE versus Central Treatment for Removal of Organic
Chemicals by Granular Activated Carbon

	Cost ($) per Household per Year					
	DBCP		TCE		1,2-DCP	
Number of Households	Central	POE	Central	POE	Central	POE
10	1,325	775	1,332	815	1,356	900
15	954	775	960	815	985	900
20	760	775	766	815	790	900
25	639	775	646	815	670	900
50	380	775	385	815	410	900

NOTE: The household water usage rate is assumed to be 80 gal per person per day, with 3.3 people per household. The POE unit includes two GAC contractors with 2 cu ft of GAC in series and a design loading of 4 gal per minute per square foot. GAC replacement is assumed to occur every 1 to 2 years. For central treatment, it was assumed that GAC postcontactors would require GAC replacement every 70 to 250 days depending on the organic contaminant removed. DBCP is dibromochloropropane; TCE is trichloroethylene; 1,2-DCP is 1,2-dichloropropane.

SOURCE: Reprinted, with permission, from Goodrich et al. (1992). ©1992 by the *Journal of the American Water Works Association.*

shown in this figure, if 20 homes are involved and the length of distribution pipe required is less than 4,000 ft. (1,200 ms), then connecting to a central treatment plant is more cost effective than using POE devices.

Use of POE and POU treatment devices to satisfy drinking water regulatory requirements may be appropriate in some instances, especially for very small systems. In some cases, POE might be the only affordable solution for a very small community with limited financial resources. However, the objections to using POE and POU treatment devices are substantial and have merit, particularly as the system size increases and the complexity of monitoring and servicing the devices increases. Using centralized water treatment should be the preferred option for very small systems, and POE or POU treatment should be considered only if centralized treatment is not possible.

Bottled Water Distribution

Bottled water use in the United States has increased at a rate of approximately 15 to 20 percent per year over the past 20 years (Richardson, 1991). This

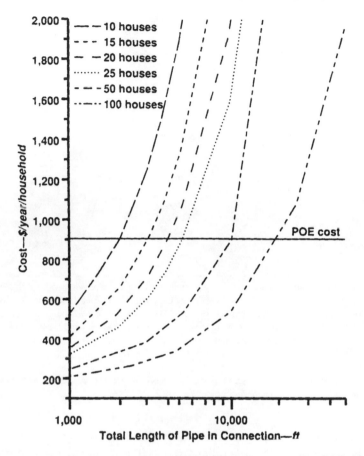

FIGURE 3-11 Cost of POE versus connecting to a central system. The POE device in this example is like that described in Table 3-5. The central treatment alternative assumes that a 6×10^3 m³/d (1.6 mgd) conventional plant serving 10,000 people exists nearby and can deliver water at $1.70 per 3,800 liters (1,000 gal). The example assumes that the conventional plant does not need any process modifications. The additional distribution system required is assumed to be a combination of 15- and 20-m (6- and 8-in.) ductile-iron pipes, fittings, and valves. SOURCE: Reprinted, with permission, from Goodrich et al. (1992). ©1992 by the *Journal of the American Water Works Association.*

increase has occurred despite the high costs of bottled water: the U.S. General Accounting Office found that "consumers may be paying as much as 300 to 1,200 times more per gallon for bottled water than for tap water because they believe it tastes better, is safe and healthy, or is free of contaminants" (Community Nutrition Institute, 1991). The majority of bottled water is purchased for aesthetic reasons rather than for quality reasons related to drinking water regulations.

Some bottled water is used by necessity rather than because of personal preferences. Examples of necessary uses include water used in areas that have experienced floods, earthquakes, or hurricanes. Bottled water is commonly provided to those who cannot boil water, such as motel and hotel patrons, when a community experiences a waterborne disease outbreak. Bottled water is now recognized as an alternative water supply for emergency purposes by the Department of Interior's Emergency Water Supply Plan, the U.S. Army Corps of Engineers' Emergency Water Plan, and the EPA's National Contingency Plan under the Superfund act. In addition, the EPA rules specify that bottled water, like POU devices, may be used on a temporary basis to avoid anunreasonable risk to health or as a condition of a variance or exemption to drinking water regulations.

Bottled water comes from a variety of sources, including springs, artesian wells, and even public water systems. Bottled water derived from municipal water systems may be treated with ozone and GAC to enhance its taste and odor properties before it is bottled. The Food and Drug Administration (FDA) regulates bottled water. However, the FDA regulates fewer contaminants than does the EPA under the SDWA. If bottled water were to be provided to customers of a small water system as a means of meeting EPA regulations, bottlers who use public water supplies as their sources would probably be appropriate choices to consider, as the status of compliance with EPA regulations for the source of the bottled water would be known or readily available.

Distribution of bottled water is an important issue to resolve if a small system uses bottled water to comply with EPA regulations. One approach would be to have a supply available at the town hall or the water system office for water system customers to take home at no charge. Another approach would be to deliver a supply of bottled water to each household on a regular basis. In a recent American Water Works Association (AWWA) Research Foundation project, a supply of bottled water was delivered once every 2 weeks to each family participating in a study involving bottled water (R. Karlin, AWWA Research Foundation, personal communication, 1996). If more water was needed before the end of the 2 weeks, study participants called and more water was provided. Because of the logistics of providing bottled water, it is appropriate only for intermittent or short-term purposes, rather than for continuous, long-term needs.

CONCLUSIONS

The complexity of choosing, financing, operating, and maintaining a small water supply system cannot be overstated. Technology applications differ in their suitability for different water sources and water system sizes. Important factors in choosing a treatment technology for the small water supply system include regulatory compliance; source water quality; capital, operational, and maintenance expenses; and expertise required to operate the system.

In selecting drinking water treatment technologies, small communities should keep the following considerations in mind:

• **Small systems should apply technologies to meet requirements of the Safe Drinking Water Act only after exhausting all other possible options.** Other routes to compliance include finding an alternative water source, linking with another water system, or purchasing treated water from another system.

• **No single water treatment process can solve all water quality problems.** Water systems may need to apply a sequence of technologies to meet all regulatory requirements and customer preferences.

• **The most cost-effective way to reduce the incidence of most types of waterborne disease caused by microbial pathogens is to disinfect the water.** Free chlorine is the easiest type of disinfectant for small systems to apply to meet requirements of the SDWA. However, other strategies, such as use of ozone prior to treatment followed by use of chloramine in the water distribution system, may be needed to minimize the formation of disinfection byproducts that are already or will soon be regulated.

• **For small systems using ground water sources, the most commonly reported chemical contaminants influencing the selection of water treatment systems are nitrate, fluoride, and volatile organic compounds.** Elevated nitrate and fluoride levels can be reduced with ion exchange, electrodialysis reversal, or reverse osmosis systems. Volatile organic compounds can be stripped from the water by aeration. Other types of synthetic organic compounds can be treated by adsorption on granular or powdered activated carbon.

• **For small systems using surface water sources, treatment requirements are driven by the Surface Water Treatment Rule, which requires filtration and disinfection of the water.** Membrane filtration systems may best address the variety of problems in surface water because they simultaneously remove microbial contaminants (although disinfection is still required), organic matter that can form disinfection byproducts, and, in the case of reverse osmosis, inorganic chemicals. Slow sand filtration is an appropriate treatment process for surface waters of high quality.

• **Automated devices for monitoring small water systems can allow several small systems to share an operator, who can be better trained than a part-time operator.** However, remote monitoring does not eliminate the need for routine maintenance checks.

• **Very small water systems (those serving fewer than 500 people) may consider using point-of-use or point-of-entry treatment devices in individual homes as an alternative to centralized treatment if all other options are too costly.** However, maintenance and compliance responsibilities must remain with the water supplier rather than with the individual homeowner. Developing institutional arrangements for managing these systems may be a greater challenge than finding technology that is effective for removing the contaminants of concern and may elevate the costs of these units above the costs of central treatment. In the case of POU devices, the need to enter customers' homes to service the equipment, plus the fact that these devices treat water at only one tap, may preclude their use as a long-term solution to water quality problems.

• **Bottled water can be an acceptable short-term solution for providing drinking water of acceptable quality.** However, because of the difficulties associated with distributing it and making sure consumers do not ingest the tap water, it is not an appropriate long-term solution.

REFERENCES

AWWA (American Water Works Association. 1980. The Status of Direct Filtration. Journal of the American Water Works Association 72(7):405–411.

AWWA Committee. 1992. Survey of water utility disinfection practices. Journal of the American Water Works Association 84(9):121–128.

AWWA. 1995. Electrodialysis and Electrodialysis Reversal, First Edition. Denver: AWWA.

AWWA Research Foundation. 1996. Internal Corrosion of Water Distribution Systems, Second Edition. Denver: AWWARF.

Bauer, M. J., J. S. Colbourne, D. M. Foster, N. V. Goodman, and A. J. Rachwal. 1996. GAC enhanced slow sand filtration. Pp. 223-232 in Advances in Slow Sand Filtration and Alternative Biological Filtration, N. Graham and R. Collins, eds. New York: John Wiley & Sons.

Bellamy, W. D., D. W. Hendricks, and G. S. Logsdon. 1985a. Slow sand filtration: Influences of selected process variables. Journal of the American Water Works Association 77(12):62–66.

Bellamy, W. D., G. P. Silverman, D. W. Hendricks, and G. S. Logsdon. 1985b. Removing *Giardia* cysts with slow sand filtration. Journal of the American Water Works Association 77(2):52–60.

Bellen, G., M. Anderson, and R. Gottler. 1985. Management of Point-of-Use Drinking Water Treatment Systems. Ann Arbor, Mich.: National Sanitation Foundation.

Benjamin, L., R. W. Green, A. Smith, and S. Summerer. 1992. Pilot testing a limestone contractor in British Columbia. Journal of the American Water Works Association 84(5):70–79.

BETZ. 1980. BETZ Handbook of Industrial Water Conditioning. Trevose, Pa.: BETZ Laboratories, Inc.

Brigano, F. A., J. P. McFarland, P. E. Shanaghan, and P. Burton. 1994. Dual-stage filtration proves cost-effective. Journal of the American Water Works Association 86(5):75–88.

Cheryan, M. 1986. Ultrafiltration Handbook. Lancaster, Pa.: Technomic Publishing.

Cleasby, J. L. 1991. Source water quality and pretreatment options for slow sand filters. Chapter 3 in Slow Sand Filtration, G. S. Logsdon, ed. New York: American Society of Civil Engineers.

Cleasby, J. L., D. J. Hilmoe, and C. J. Dimitracopoulos. 1984. Slow sand and direct in-line filtration of a surface water. Journal of the American Water Works Association 76(12):44–55.

Community Nutrition Institute. 1991. FDA not enforcing rules on bottled water: GAO. Nutrition Week (April):6.

Conlon, W. J. 1990. Membrane processes. Chapter 11 in Water Quality and Treatment: A Handbook of Community Water Supplies, Fourth Edition. Denver: American Water Works Association.

Cullen, T. R., and R. D. Letterman. 1985. The effect of slow sand filter maintenance on water quality. Journal of the American Water Works Association 77(12):48–55.

Duranceau, S. J., J. S. Taylor, and L. A. Mulford. 1992. SOC removal in a membrane softening process. Journal of the American Water Works Association 84(1):68–78.

Edwards, M., M. R. Schock, and T. B. Meyer. 1996. Alkalinity, pH, and copper corrosion by-product release. Journal of the American Water Works Association 88(3):81–94.

EPA (Environmental Protection Agency). 1982. Memorandum from Joseph A. Cotruvo, Director, Criteria and Standards Division, ODW, to Greene A. Jones, March 3.

EPA. 1985. Federal Register 50(219):46880.

EPA. 1989. Federal Register 40 CFR Parts 141 and 142, Thursday, June 29.

EPA. 1994. The National Public Water System Supervision Program: FY 1993 Compliance Report. EPA 812-R-94-001. Washington, D.C.: Office of Water.

EPA. 1995. Letter from Jennifer Orme Zavaleta, Acting Chief, Human Risk Assessment Branch, Offices of Sciences and Technology, to S.K. Verma, February 14.

Fox, K. R., R. J. Miltner, G. S. Logsdon, D. L. Dicks, and L. F. Drolet. 1984. Pilot-plant studies of slow-rate filtration. Journal of the American Water Works Association 76(12):62–68.

Goodrich, J. A., J. Q. Adams, B. W. Lykins, and R. M. Clark. 1992. Safe drinking water from small systems: treatment options. Journal of the American Water Works Association 84(5):49–55.

Grubbs, T. R., and F. W. Pontius. 1992. U.S. EPA releases draft ground water disinfection rule. Journal of the American Water Works Association 84(9):25–31.

Haarhoff, J., and J. L. Cleasby. 1991. Biological and physical mechanisms in slow sand filtration. Chapter 2 in Slow Sand Filtration, G. S. Logsdon, ed. New York: American Society of Civil Engineers.

Hall, T., and R. A. Hyde. 1987. Evaluation of Slow Sand Filtration for Small Supplies: Package Plant Evaluation at Stalling Busk. Final Report. Stevenage, England: Water Research Centre.

Hall, T., J. Pressdee, R. Gregory, and K. Murray. 1994. Cryptosporidium removal during water treatment using dissolved air flotation. Pp. 110-114 in Flotation Processes in Water and Sludge Treatment, Proceedings of International Association of Water Quality—International Water Supply Association—American Water Works Association Joint Specialized Conference. Orlando, Fla.: Chameleon Press Ltd.

JAWWA (Journal of the American Water Works Association). 1993. Reverse osmosis solves radium problem. Journal of the American Water Works Association 85(6):111–112.

Kollajtis, J. A. 1991. Dissolved air flotation applied in drinking water clarification. Pp. 433-448 in Proceedings 1991 AWWA Annual Conference. Denver: American Water Works Association.

Letterman, R. D., and G. S. Logsdon. 1976. Survey of direct filtration practice—Preliminary report. Paper Presented at AWWA Annual Conference, New Orleans, La., June 20-25, 1976.

Letterman, R. D., C. T. Driscoll, M. Haddad, and H. A. Hsu. 1987. Limestone Bed Contractors for Control of Corrosion at Small Water Utilities. EPA/600/2-86/099. Cincinnati: Environmental Protection Agency.

Letterman, R. D., M. Haddad, and C. T. Driscoll. 1991. Limestone contractors: Steady-state design relationships. Journal of Environmental Engineering 117(3):339-358.

Logsdon, G. S., M. M. Frey, T. D. Stefanich, S. L. Johnson, D. E. Feely, J. B. Rose, and M. Sobsey. 1994. The Removal and Disinfection Efficiency of Lime Softening Processes for *Giardia* and Viruses. Denver: American Water Works Association Research Foundation.

Lundelius, E. F. 1920. Adsorption and solubility. Kolloid-Zeitschrift 26:145-151.

Lykins, W., Jr., R. M. Clark, and J. A. Goodrich. 1992. Point-of-Use/Point-of-Entry for Drinking Water Treatment. Boca Raton, Fla.: Lewis Publishers, Inc.

Macler, B. A. 1996. Developing the ground water disinfection rule. Journal of the American Water Works Association 88(3):47–55.

Malley, J. P., Jr., and J. K. Edzwald. 1991. Laboratory comparison of DAF with conventional treatment. Journal of the American Water Works Association 83(9):56–61.

McKelvey, G. A., M. A. Thompson, and M. J. Parrotta. 1993. Ion exchange: A cost-effective alternative for reducing radium. Journal of the American Water Works Association 85(6):61–66.

Meller, F. H.   1984.   Electrodialysis (ED) and Electrodialysis Reversal (EDR) Technology. Watertown, Mass.: Ionics, Inc.

Montgomery, J. M., Consulting Engineers.   1985.   Water Treatment Principles and Design.   New York: John Wiley and Sons.

Morin, O. J.   1994.   Membrane plants in North America.   Journal of the American Water Works Association 86(12):42–54.

Opferman, D. J., S. G. Buchberger, and D. J. Arduini.   1995.   Complying with the SWTR: Ohio's experience.   Journal of the American Water Works Association 87(2):59–67.

Poynter, S. F. B., and J. S. Slade.   1977.   The removal of viruses by slow sand filtration.   Progress in Water Technology 9(1):75-88.

Pyper, G. R.   1985.   Slow Sand Filter and Package Treatment Plant Evaluation: Operating Costs and Removal of Bacteria, Giardia, and Trihalomethanes.   EPA/600/2-85/052.   Cincinnati: Environmental Protection Agency.

Pyper, G. R., and G. S. Logsdon.   1991.   Slow sand filter design.   Chapter 5 in Slow Sand Filtration, G. S. Logsdon, ed.   New York: American Society of Civil Engineers.

Richardson, S. E.   1991.   Reporting to Congress on California's bottled water program.   Journal of the Association of Food and Drug Officials 55(2):45–49.

Riesenberg, F., B. B. Walters, A. Steele, and R. A. Ryders.   1995.   Slow sand filters for a small water system.   Journal of the American Water Works Association 87(11):48–56.

Robeck, G. G., N. A. Clarke, and K. A. Dostal.   1962.   Effectiveness of water treatment processes for virus removal.   Journal of the American Water Works Association 54(10):1275–1290.

Schlenger, D. L., W. F. Riddle, B. K. Luck, and M. H. Winter.   1994.   Automation Management Strategies For Water Treatment Facilities.   Denver: American Water Works Association Research Foundation and American Water Works Association.

Schuler, P. F., M. M. Ghosh, and S. N. Boutros.   1988.   Comparing the removal of Giardia and Cryptosporidium using slow sand and diatomaceous earth filtration, Pp. 789-805 in Proceedings 1988 AWWA Annual Conference.   Denver: American Water Works Association.

Seelaus, T. J., D. W. Hendricks, and B. A. Janonis.   1986.   Design and operation of a slow sand filter.   Journal of the American Water Works Association 78(12):35–41.

Semmens, M. J., R. Qin, and A. K. Zander.   1989.   Using a microporous hollow-fiber membrane to separate VOCs from water.   Journal of the American Water Works Associaiton 81(4):162–167.

Snoeyink, V. L.   1990.   Adsorption of organic compounds.   Chapter 13 in Water Quality and Treatment: A Handbook of Community Water Supplies, Fourth Edition.   Denver: American Water Works Association.

Sorg, T. J.   1978.   Treatment technology to meet the interim primary drinking water regulations for inorganics.   Journal of the American Water Works Association 70(2):105–112.

Sorg, T. J., R. W. Forbes, and D. S. Chambers.   1980.   Removal of Radium-226 from Sarasota County, Fla. drinking water by reverse osmosis.   Journal of the American Water Works Association 72(4):230–237.

Syrotynski, S., and D. Stone.   1975.   Microscreening and diatomite filtration.   Journal of the American Water Works Association 67(10):545–548.

Tamburini, J. U., and W. L. Habenicht.   1992.   Volunteers integral to small system's success.   Journal of the American Water Works Association 84(5):56–61.

Weber, W. J., Jr.   1972.   Physicochemical Processes for Water Quality Control.   New York: Wiley-Interscience.

Wiesner, M. R., J. Hackney, S. Sethi, J. G. Jacangelo, and J. M. Laine.   1994.   Cost estimates for membrane filtration and conventional treatment.   Journal of the American Water Works Association 86(12):33–41.

Zander, A. K., M. J. Semmens, and R. M. Narbaitz.   1989.   Removing VOCs by membrane stripping.   Journal of the American Water Works Association 81(11):76–81.

4

Evaluating Technologies for Small Systems

Before installing a new water treatment system, water utilities must obtain approval from state drinking water regulators. Prior to granting approval, regulators may require pilot tests, depending on the technology to be installed. Package water treatment plants often use innovative designs to fit the treatment processes into compact units, and therefore regulators are often hesitant to approve them without detailed pilot testing. Pilot tests can last for periods of time as short as several weeks or as long as 1 year or more. Long programs of pilot testing add substantially to the costs of installing package plants for small systems. For example, one equipment manufacturer reported that pilot testing increased the capital cost of a treatment system by 28 percent (McCarthy, 1995).

This chapter discusses the degree to which testing of treatment technologies appropriate for small communities can be standardized. It describes when preexisting pilot test data or plant operating data are adequate to ensure performance of the technology at a new location and when site-specific testing is necessary. The chapter also discusses the availability of data on performance of water treatment technologies for small systems.

Before making decisions about treatment processes to employ and the extent of pilot testing that will be necessary, water system engineers will need to obtain information on raw water quality and desired treated water quality. Safe Drinking Water Act (SDWA) regulations specify the basic requirements for finished water quality. Because of customer or management preferences, water systems may also decide to add additional treatment (for example, water softening) not required under the SDWA. In some situations, the source water quality may be so high that the water meets all SDWA requirements and customer demands without treatment. Where treatment is needed, in some cases water system engineers can select a treatment system based on performance data from other locations with water of similar quality or based on relatively inexpensive bench-scale tests. In other situations, however, available information on source water quality may provide convincing evidence that pilot testing for a particular treatment process is needed before a full-scale plant is installed. Thus, while current requirements for pilot testing of water treatment technologies lead to some duplication of effort and can be reduced, for certain combinations of treatment technologies and source waters, some degree of site-specific pilot testing always will be necessary to ensure that the equipment will perform adequately.

CURRENT REQUIREMENTS FOR PILOT TESTING

Regulators require pilot tests in part to ensure that the water treatment system, whether package or custom designed, will effectively treat the water at the particular location. They regard such testing as especially important for surface water systems, for which water quality can be highly variable not only from place to place but also from season to season. For example, filter-clogging algae can appear in surface waters intermittently.

Because uncontrollable factors such as nutrient matter in the water and sunlight strongly influence algae growth, prediction and control of algae blooms is difficult if not impossible in the context of small system operations. Similarly, turbidity in some reservoirs and in many rivers varies for reasons—such as heavy rainfall and runoff, flooding, and heavy runoff from melted snow—that water utilities cannot control. Ground water tends to have more consistent quality than surface water, so, theoretically, site-specific testing is less important for ground water systems when performance data are available from other locations. However, even where the quality of the source water is relatively high and consistent, regulators usually require that package plants be pilot tested because of concern about the legitimacy of performance data provided by the manufacturers, which the regulators may perceive as a "sales pitch" (GAO, 1994). Regulators have indicated that independent, third-party evaluations of package devices are lacking, so the performance of package plants usually must be verified at each new location even when the manufacturer claims to have used the technology on water of similar quality elsewhere (GAO, 1994).

Requirements for pilot tests may vary significantly from state to state and even within a given state (WMA, 1994). For example, Illinois regulators always require a pilot study, usually lasting three seasons, for systems that treat surface water because of the high variability of surface water quality; for ground water systems they almost always require a 3- to 4-week pilot test. Similarly, New York regulators almost always require site-specific pilot studies prior to approving package systems. In Minnesota, conversely, regulators will approve package plants without pilot testing if the plants have been proven effective on waters of similar quality at other locations. In Pennsylvania, engineers in six regional offices decide on the extent of pilot testing, and pilot testing requirements therefore vary within the state.

Pilot data collected in one state may not be considered valid in other states. For example, one equipment manufacturer reported that several states have refused to approve a technology that has operated effectively at more than 100 sites nationwide because of their reluctance to use data from other states (GAO, 1994). Seven western states (Alaska, California, Colorado, Idaho, Montana, Oregon, and Washington) attempted to encourage information sharing and to streamline their testing requirements for filtration systems, including package technologies, by developing a guidance document known as the Western States Protocol (GAO, 1994; David Clark, Washington State Department of Health, personal communication, 1995). The protocol applies to technologies not defined as "conventional" in the Surface Water Treatment Rule (SWTR), the U.S. Environmental Protection Agency (EPA) regulation that specifies filtration requirements for surface water. Examples of package technologies that could be evaluated under the Western States Protocol are bag filters and cartridge filters. However, rather than uniformly implementing the protocol, individual states have modified it to meet their specific needs. Although modification of a testing protocol perhaps could be justified on the basis of individual needs that vary from state to state, the overall effect of such modifications is to make the transfer of accumulated testing data from state to state difficult. An analogous situation would be if modifications were permitted in a standard EPA method for testing water quality, resulting in each state using a slightly different analytical approach for measuring water quality parameters such as turbidity and free chlorine concentration.

ESTABLISHING A THIRD-PARTY CERTIFICATION PROGRAM

State regulators would likely reduce requirements for extensive piloting of package technologies on a case-by-case basis if equipment manufacturers could receive credible third-party certification of their products. A third party is a technically and otherwise competent body other than one controlled by the producer or buyer. Certification would provide assurance to decisionmakers that the product is capable of performing as advertised. (Of course, certification would not release the system from operation and maintenance activities to keep the equipment performing properly.)

Under a third-party certification program, manufacturers would voluntarily submit their equipment or processes to a certification agency for approval. Certification would include three key elements:

 • verification of the manufacturer's claims, especially claims of reductions in contaminant levels;
 • testing of construction materials to ensure that they are safe for contact

---

BOX 4-1 NSF Standards for Water Treatment Equipment

NSF International has limited standards for point-of-use and point-of-entry devices, drinking water treatment chemicals, and drinking water system components but none (yet) that specifically apply to individual package treatment systems. Two of the standards cover the ability of point-of-use and point-of-entry devices to improve aesthetic properties of the water and eliminate compounds that cause adverse health effects (see Chapter 3). Four cover specific point-of-use and point-of-entry devices for individual home use: ultraviolet systems, reverse osmosis systems, distillation systems, and cation exchange systems (McClelland, 1994). NSF standards for water treatment chemical additives and drinking water system components are aimed at ensuring that the treatment process itself does not create health hazards in the drinking water.

NSF uses expert committees to develop its technology standards. The primary committee developing a standard, known as the "joint committee," includes representation from industry, government, and consumer groups. This committee also receives input from a council of public health consultants and a certification council that has expertise in test methods. Once an NSF committee develops a standard, the NSF applies to have it certified by the American National Standards Institute (ANSI). An ANSI designation means that only one standard exists for that type of product in the United States and that the standard follows all of ANSI's guidelines.

Point-of-use and point-of-entry devices undergoing NSF review must meet requirements in four basic areas: (1) the equipment must meet manufacturers' claims for the level of contaminant removal provided; (2) the materials used in the equipment must be disclosed and tested for leachability; (3) the equipment must be tested for structural soundness; and (4) the equipment must have an adequate installation manual and must be accurately labeled.

---

with drinking water and are capable of handling operating conditions for the expected life of the equipment; and

• evaluation of operation and maintenance manuals to ensure that they provide accurate and complete information about the equipment.

Third-party certification is currently available through the National Sanitation Foundation (NSF) International for a limited number of point-of-use and point-of-entry treatment devices and for certain water treatment chemicals and system components (see Box 4-1). However, certification is not yet available for package plants. In late 1995, the EPA launched a program, the Environmental Technology Verification Program, to test a wide range of environmental technologies, including package water treatment plants. The EPA has provided funding for NSF International to develop equipment performance verification protocols and test plans for evaluation of water treatment package plants. A key aspect of the program is the involvement of state regulators in protocol development. After the EPA and NSF International develop the protocols, third parties will test the technologies in a manner intended to facilitate acceptance by state regulatory agencies while reducing the burden of repeated testing now faced by equipment manufacturers and vendors. The EPA funded the program with the intention of continuing support for 3 years, with hopes that testing fees will sustain most of the costs after this period and that testing will continue as manufacturers and entrepreneurs develop new treatment technologies.

PROTOCOLS FOR TECHNOLOGY TESTING: PRINCIPLES

Standard protocols for testing water treatment equipment exist for a variety of technologies, but they have not been collected in a common location. As a result, designers of water treatment systems conduct bench and pilot studies using their own individual methods. Whereas one experimenter may test for treatment efficiency using turbidity measurements, another may use a particle counter. These data cannot be directly compared, so they are essentially lost to the drinking water field after their initial use. Testing protocols specifying tests to perform, analytical procedures to follow, and a standard method for data reporting are essential for allowing interpretation and comparison of testing results from various sources and for eliminating unnecessary duplication in pilot testing. Whether testing is performed by the manufacturer, an engineering firm, a utility, or an approved third party, use of a standard testing protocol will aid in interpretation and acceptance of the data.

Protocols for technology testing need to be sufficiently comprehensive to ensure that the total needs of small water systems are considered when testing is carried out. The obvious requirement for water treatment processes is to produce a water quality that meets SDWA requirements and customer preferences. In addition, water systems need to look beyond the general capabilities of processes for contaminant removal and also consider process efficiency, operating ease, and operation and maintenance expenses. Installation of technology that is too complex to operate or too expensive to maintain is likely to result in regulatory compliance problems in the long run. Process efficiency and operating ease are related to factors such as length of filter run, extent of on-site operator attention needed, and extent of pretreatment required. Operation and maintenance expenses are influenced by factors such as chemical dosages, volume of water treated before system components need replacement, and energy requirements. Data on such aspects of water treatment need to be included in testing protocols.

Testing should be performed in a laboratory or pilot facility owned by the manufacturer and certified by a qualified third party or owned by a third party and having a national laboratory certification. The certification should come from the EPA, NSF International, American National Standards Institute, or another nationally recognized organization. National licensing of the laboratory is necessary to ensure that state regulators, and ultimately the organization that certifies the technology, will accept the data.

Tests must be performed at least in duplicate for each operating condition.

All analytical methods should follow a nationally accepted format such as those outlined in *Standard Methods for the Examination of Water and Wastewater* (APHA, AWWA, and WEF, 1992). Testers should measure and report initial and final concentrations for every regulated contaminant for which the equipment manufacturer claims a concentration reduction.

Evaluators should test the systems at the low end of the expected water temperature range and also at the high end if a high temperature can have any adverse effect. They should also perform tests at the high and low end of the recommended operating pressure and flow range. The protocol should require reporting of the full range of data collected.

Information should be obtained on raw water quality or on water quality before some process modification and on treated water quality or the quality of the water after some process modification. During the study, information on operating parameters such as water flow rate, chemical feed rates, operating pressure, contact time, filter bed lead loss, mixer or flocculator speed, backwash cycle time, empty bed contact time, air flow rate, percent of water rejected by a membrane, and so on should also be collected. As a general principle, if an adjustment can be made to process equipment to change its operation, that aspect of the equipment operation should be noted, and it might become a variable in the testing program. If equipment operation changes but the tester fails to note such changes, performance might seem erratic without any obvious reason.

Numerous kinds of miscellaneous data may be of use. Weather observations, river flows, lake or reservoir levels, industrial or municipal waste discharges, changes in equipment at the treatment plant, or repairs and maintenance in the distribution system might have impacts on various sorts of pilot studies. A good rule of thumb is to write down any observation that might be of value later. This at first seems burdensome, but thoroughly collected data can be used later to explain results, and if no problems develop in testing then some observations may prove unnecessary and would not have to be archived. Equipment performance data, void of supporting background information, are not likely to be accepted for waters other than the source water involved in the test program.

Finally, the protocol should require that results be measured and reported directly in the units specified in the regulatory language. For example, a particle-counting measurement cannot substitute for a turbidity measurement. Thus, the common language of testing protocols must be the language of the drinking water regulations.

TECHNOLOGY-SPECIFIC TESTING REQUIREMENTS

The degree to which test results from one location can be applied elsewhere, and therefore the extent to which third-party certification can reduce piloting requirements, depends on the technology and source water. Some types of equipment, such as chemical feeders, can be designed and tested, and all chemical feeders of identical size and design should perform the same under identical equipment operating conditions. Similarly, aeration systems, disinfection processes, bag filters, and cartridge filters rarely require pilot testing. Performance of other types of equipment, such as a package plant employing coagulation and filtration, adsorption, membrane filtration, or lime softening, depends on the quality of the source water. For such systems, some degree of site-specific evaluation may be essential, as explained in detail below. Site-specific testing can range from bench-scale evaluations to operation of a pilot plant for a given time period. The extent of site-specific testing required depends on the amount of data available for the same or similar source waters.

Aeration Systems

The performance of aeration systems can be predicted with design equations (for example, see Kavanaugh and Trussell, 1980). Therefore, pilot testing should not be required for a well-designed system. The design equations are based on data from hundreds of installations using a variety of source waters. When used with a safety factor, they allow engineers to determine without site-specific testing the size of the units needed for removal of volatile organic compounds, radon, carbon dioxide, and natural compounds that cause taste and odor problems. Designers must pay careful attention to possible foulants, such as reduced iron and manganese, commonly found in ground water.

Membrane Processes

Several approaches are possible for evaluating membrane processes. Which approach is appropriate depends on the nature of the source water and the principles by which the particular membrane removes contaminants from water.

For ground water systems using membranes, the only site-specific analysis that may be necessary is evaluation of source water characteristics to determine the potential for chemical scaling of the membranes. Scaling is especially a concern for brackish ground water.

Assessing the treatability of surface waters by membranes is somewhat more complex. The capability of membrane processes to cope with particulate matter in raw water is limited. Thus bench-scale testing may be necessary to determine whether pretreatment is required to remove a portion of the particulate matter.

Currently, researchers recommend a pilot study on a single membrane element for 1,000 hours in each of the wet and dry seasons when membranes are used to treat surface waters (Taylor and Mulford, 1995). As more documentation is developed on membrane process capabilities, the need for pilot testing should decline. For example, for microfiltration, certain types of microbiological contaminants are too large to pass through the membrane pores, and therefore, once the removal capability has been established it should theoretically be unnecessary to demonstrate the technology at every new application site. Similarly, once tests have proven the ability of a particular reverse osmosis membrane to reject an inorganic contaminant such as uranium or radium, retesting the same membrane at many different locations to verify the uranium or radium removal capability should not be required. This same concept applies to the other membrane processes (ultrafiltration, nanofiltration, and electrodialysis/electrodialysis reversal) when used to remove contaminants for which their effectiveness has been documented, although some testing specific to the source water will be necessary to determine the potential for membrane fouling.

Pilot testing requirements for membranes used in surface water treatment are already being reduced in some states. For example, based on testing by the Metropolitan Water District of Southern California at one site on the Colorado River Aqueduct (Kostelecky et al., 1995), the California Department of Health Services approved use of the Memcor microfiltration process for meeting requirements of the SWTR at several other sites along the aqueduct. The department granted 3-log removal credit for *Giardia* cysts (meaning regulators will assume the membrane can remove 99.9 percent of these organisms) and 0.5-log (meaning 68 percent) removal credit for viruses.

Adsorption Processes

Bench-scale testing of the effectiveness of the particular adsorptive material (activated carbon, ion exchange, or activated alumina) for the target contaminant or contaminants in the raw water is the minimum level of evaluation necessary to design an adsorption system.

For granular activated carbon (GAC) systems, the purpose for which the system is employed (adsorption of dissolved organic compounds, biological stabilization of the water, or particle removal) will influence the type of site-specific testing necessary to design a full-scale system. The most accurate technique for predicting the performance of a full-scale GAC system is a pilot-scale system using the same source water. A pilot system treats the same specific flow rate, carbon size, and influent water as the planned full-scale column but uses a smaller diameter and thus less carbon. Contaminant breakthrough in pilot columns has been shown to closely model full-scale breakthrough (Oxenford and Lykins, 1991). Pilot testing is expensive and time-consuming, however. Although not effective for testing biological or particle filtration applications, small-scale columns can accurately predict the performance of a full-scale GAC system for many adsorption applications (Crittenden et al., 1991). In small-scale tests, the GAC is crushed until the grain-size:column-size ratio is equivalent to that of the full-scale system. The crushed carbon is installed in a column that may be only a couple of inches in length. This small-scale column is used to evaluate the performance of the carbon in treating the source water. Small-scale column tests are rapid and inexpensive and greatly reduce the time and money required to size a full-scale column.

As with GAC systems, ion exchange and activated alumina bench- or pilot-scale testing can provide the information needed for full-scale design of a system. It is important that the tests, no matter what scale, are performed on the source water to take into account competitive adsorption from other ionic species present.

Powdered activated carbon (PAC) addition to a mixed tank is a relatively inexpensive method of reducing organic contaminant concentrations in finished water. Equilibrium isotherm models, however, do not reliably predict PAC performance in water because contaminant characteristics and the effectiveness of mixing have a strong effect on the amount of contaminant removed by PAC. In addition, as for GAC, background organic compounds affect PAC performance. For these reasons, testing must involve the actual source water. Bench-scale testing can be accurate if careful attention is given to reproducing the mixing characteristics of the full-scale system. However, since existing pilot data on the same source would likely not mimic the intended full-scale mixing at a new site, adsorption kinetics must be taken into account when using such data.

Coagulation/Filtration Systems
(Conventional Filtration, Direct Filtration, and Dissolved Air Flotation)

Because the physical and chemical principles governing the performance of coagulation/filtration systems are so complex, some degree of site-specific testing will always be necessary for these technologies unless the technology has proven effective at a different installation using the same source water. In some cases, bench-scale tests using jars to determine appropriate coagulant doses will be adequate, but in other cases site-specific pilot tests will be necessary. The degree of testing required depends in part on the design of the coagulation/filtration system and in part on the characteristics of the raw water.

Effect of Filtration System Design on Testing Requirements

A considerable experience base exists on the ability of conventional treatment trains (coagulation followed by flocculation, sedimentation, and filtration) to successfully treat a broad range of water quality. Therefore, site-specific testing requirements are less extensive for package plants employing conventional treatment trains than for those using newer technologies. When a package plant employs newer methods, such as upflow or downflow granular media beds to flocculate and remove particles before filtration, site-specific pilot testing is likely to be needed unless the range of raw water quality characteristics is well within the values for which the equipment has been demonstrated to the satisfaction of consultants and regulatory engineers. Direct filtration plants should always be pilot tested unless one is already operating successfully to treat the same source of water because these systems are so highly sensitive to water quality.

Through a centralized pilot testing program, it may be possible to reduce site-specific pilot testing requirements for the various kinds of filtration technologies, especially those for which the experience base is not extensive. The range of turbidity that the filter can manage could be determined by testing a very muddy source water and dilutions of that water to provide a range of raw water turbidities for evaluation. Filtered water turbidity and rate of head loss development would be the key performance parameters to document in such testing. Similar tests could be carried out on waters having a wide range of color. The objective would be to treat water of higher and higher turbidity or color to the point of reaching either failure of the process or a very high upper limit, such as 2,000 nephelometric turbidity units or 400 to 500 color units. Finding a quality of water that was not treatable, or documenting the capability of a process train to treat raw water worse than would be encountered in an extreme case, would provide a sound basis for defining the appropriate water quality limits or for determining that nearly any expected raw water turbidity or color could be treated.

Performance limits for package filtration systems using unconventional technologies could also be established by evaluating the effectiveness of existing installations. Manufacturers could provide lists of each installation of their equipment to an appropriate neutral body, which could then review the data to assess the range of raw water quality characteristics that the filter can manage (see "Centralizing Data Collection" later in this chapter). In particular, dissolved air flotation, although used extensively in Europe and South Africa, is rarely applied in the United States. Consultants and regulatory engineers would benefit from the availability of more performance data that delineates the ability of this process to handle raw water turbidity.

Effect of Raw Water Quality on Testing Requirements

Regardless of the type of coagulation/filtration technology, some level of site-specific testing will always be required, at a minimum to determine appropriate coagulant doses, unless the identical system is treating water from the same source at another facility. Whether bench-scale testing will be sufficient or more extensive pilot tests will be required depends in large part on the source water quality. Source waters with a single quality factor that needs to be treated present the simplest cases and require the least amount of prior testing and evaluation. Examples of these are waters that have no algae and either high turbidity but low color or low turbidity but high color. When only a single problem needs to be evaluated, determining the appropriate coagulant dose is much less difficult than when multiple factors, such as high turbidity and high color, are involved. Conversely, for waters with various combinations of turbidity, color, and algae, site-specific testing is unavoidable for coagulation/filtration technologies. Tables 4-

1 and 4-2 show the types of site-specific pilot data that might be collected during pilot testing of a conventional coagulation/filtration system for a source water with moderate turbidity, algae problems, color, and periodic iron and manganese. For all constituents except turbidity, water samples would be obtained for analysis during the steady-state portion of the filter run, after the initial hour or two of operation, when turbidity improves, but before the end of the run, when turbidity breakthrough (increase) might occur. Continuous measurement of filtered water turbidity provides a record of the complete filter run from beginning to end.

Testing for Removal of Turbidity. Bench-scale jar testing is sufficient for determining the performance of package plants employing coagulation and conventional filtration if the quality of water to be treated falls within the range of water quality for which the package plants have already proven effective. In jar tests, coagulant doses are determined by adding several different doses to laboratory jars containing samples of the source water, stirring the samples, and measuring the turbidity of the treated water after flocculation and sedimentation. As mentioned above, for the other types of coagulation/filtration systems some degree of pilot testing, in addition to jar testing, may be required, depending on the technology and the base of experience in using it.

Testing for Removal of Color. For removal of color with package plants using conventional filtration, jar testing followed by a brief program of pilot testing at cold temperatures and high color concentrations can sufficiently demonstrate process performance. Pilot testing of a broader range of conditions may be necessary for package coagulation/filtration plants using technologies other than conventional ones.

Testing for Removal of Turbidity and Color. Site-specific pilot testing is likely to be needed, even for conventional treatment systems, when both color and turbidity are high. Attaining effective removal of turbidity and color simultaneously can be difficult and usually requires trial-and-error testing to determine optimum coagulant doses, pH, and equipment operating parameters.

Testing for Removal of Algae. Pilot testing on site is unavoidable for source waters with algae problems unless the algae-laden water is low in turbidity, in which case dissolved air flotation would be applicable because of its proven capabilities to treat such waters. Many types of algae are filter cloggers, causing severe head loss problems unless removed ahead of the filter. Therefore, pilot testing is required to determine the chemical doses and system adjustments needed to ensure removal of the algae prior to filtration.

TABLE 4-1 Water Quality Sampling and Analysis for Testing of a Conventional Coagulation/Filtration System

Constituent	Sampling Locations	Frequency and Condition of Sampling
Temperature	R, F	Once per day
Alkalinity	R, F	Once per week and each time coagulation pH or alum dose is changed
Aluminum	R, F	Once per week when alum is used as coagulant
Iron and manganese	R, F	Once per week before reservoir turnover; after turnover, 2 or 3 times per day
True color	R, F	Once per week when true color is 5 or less; once per day when true color is greater than 5
Algae	R, F	Once per week in raw water only for monitoring; twice per day during periods of elevated algae concentration
Total organic carbon	R, F	Every week, coordinated with simulated distribution system testing for trihalomethanes and haloacetic acids
Simulated distribution system trihalomethanes	R, F	Once per month
Simulated distribution system haloacetic acids	R, F	Once per month
pH	R, C, F	Once per 8 hours
Turbidity	R, F	Measured continuously by flow-through turbidimeters; measure manually once per day to check each flow-through turbidimeter

NOTE: Sampling locations: R, raw; C, coagulated water; F, filtered water. This plan was for a 9-month study at an actual site and included a reservoir turnover period. Samples would be collected more frequently in a study of shorter duration.

TABLE 4-2 Pilot Plant Operating Data and Operator Actions for Testing of a
Conventional Coagulation/Filtration System

Operating Data	Action
Chemical feed volume	Check and record each 2 hours. Refill as needed and note volumes before and after refilling.
Raw water flow and filter flow	Check hourly; adjust when more than 10 percent above or below goal.
Filter head loss	Determine and record total head loss hourly. Record all head loss piezometers each 2 hours.
RPM of rapid mixer and flocculator	Check twice per week unless change made. Note whenever change made.

NOTE: All parameters will be checked only during times when the pilot plant is staffed.

Diatomaceous Earth Filters

Diatomaceous earth (DE) filtration is well suited to small systems because coagulation is not needed for effective removal of *Giardia* and *Cryptosporidium*. However, because DE filtration is commonly used without a clarification step ahead of filtration, source water quality limitations are somewhat stringent. Various grades of diatomaceous earth are available, ranging from coarse grades with low rates of head loss build-up to fine grades with substantial rates of head loss build-up. The finer grades of diatomaceous earth are very effective for removing turbidity as well as protozoan contaminants, but the use of such grades causes filter runs to be shorter. Pilot testing may be warranted to demonstrate the effects of using different diatomaceous earth grades, both in the context of turbidity reduction and head loss build-up. Syrotinski and Stone (1975) reported on the use of microstrainers ahead of DE filters in New York as a means of removing algae ahead of DE filtration and thus prolonging the filter runs. Although a long and comprehensive pilot testing program probably would not be needed for DE filtration, a few weeks' testing is valuable for establishing the level of turbidity in filtered water that can be attained by different grades of diatomaceous earth and for indicating the length of filter runs that might be expected with a full-scale plant. The scale of pilot testing can be modest. A DE test filter having a filter area of 0.093 m^2 (1.0 sq ft) and operated at a rate of 2.4 to 4.9 m/h (1 to 2 gpm/sq ft) is adequate to provide data for design purposes.

Slow Sand Filters

Pilot testing is always necessary for designing slow sand filters unless a slow sand filter is already treating the source water in question. Understanding of slow sand filtration technology is insufficient to allow engineers to predict what filtered water turbidity an operating slow sand filter might attain based on chemical and physical analyses of a water to be treated. Construction of a slow sand filter without pilot plant testing and without prior slow sand filter operating experience on the water source in question could result in a small water system having a new filtration plant incapable of meeting one or more drinking water standards. The nature of slow sand filtration is such that after the design parameters of plant filtration rate, bed depth, and sand size have been set, there is little a plant operator can do to improve performance of a slow sand filter that does not produce water of a satisfactory quality.

Slow sand filter pilot plant testing does not have to be expensive. Pilot plant testing has been done using manhole segments and other prefabricated cylindrical products as filter vessels. Plans and a list of materials for such a pilot filter were presented by Leland and Logsdon (1991). Slow sand filter pilot facilities operate over long periods of time—up to a year—but the level of effort can be quite low, consisting of checking head loss, flow rate, water temperature, and turbidity on a daily basis and taking samples for coliform analysis once or twice per week. Leland and Logsdon (1991) provide a recommended schedule for pilot testing of slow sand filters.

Bag Filters and Cartridge Filters

The performance of bag and cartridge filters depends on careful manufacture and use of the equipment rather than on manipulation of the water or equipment during the treatment process. Therefore, testing done in advance is a good indicator of the performance potential of this filtration equipment, and site-specific pilot testing should be unnecessary. Bag filters and cartridge filters are proprietary process equipment designed and built to the specifications of their manufacturers. The filtration occurs as water passes through the bag or cartridge inside a filter housing (pressure vessel) built by the manufacturer. When the manufacturers fabricate bags and cartridges to their specifications, and when users of the bags or cartridges apply them in the proper filter housing, use of such filters should yield reproducible results for the removal of protozoan microorganisms (mainly *Giardia* and *Cryptosporidium*). While not necessary for determining whether the filter can remove protozoan organisms, on-site testing may be useful for determining what water volume the filter can treat before becoming blinded.

Lime Softening Systems

Lime softening, as described previously, is not well suited to application in systems serving fewer than approximately 2,000 people. For small systems serving more than 2,000 people, lime softening is best suited to ground water sources, which have relatively stable water quality, rather than surface water sources, which can have quality that varies rapidly over time.

For application of lime softening to ground water, bench-scale testing with a jar test apparatus is necessary to determine appropriate process pII and the necessary quantities of lime and perhaps soda ash. Doses of these chemicals should not change greatly over time unless the ground water is subject to periodic infiltration by surface water that changes in quality. For this reason, pilot plant testing is unnecessary for lime softening of ground water that is not influenced by surface water.

If lime softening of a surface water were undertaken by a small system, the requirements of the SWTR would have to be met. For a source water having stable quality, data from other lime softening plants treating source waters of the same or poorer quality, plus jar test data on the source in question, might suffice. For source waters of variable quality, pilot testing on the water in question or operating data for a nearby full-scale plant using the same source would be preferred. Again, jar test data would be helpful for evaluating treatment options in conjunction with the other data.

Disinfection Systems

Water systems need not conduct pilot plant tests of disinfection systems that use free chlorine, chloramine, chlorine dioxide, or ozone, although limited testing may be beneficial for establishing the disinfectant demand in the presence of organic compounds, iron, or manganese, especially when ozone is used. Studies of chemical disinfectants traditionally have been carried out in centralized laboratory facilities. Regulators consider the laboratory results to be applicable to all systems. Extensive laboratory tests yield information on the extent of microorganism kill that the disinfectant can attain over a range of conditions of temperature, pH, and exposure time. From these data, the EPA has developed tables that specify the concentration and time (CT) conditions needed for inactivation of *Giardia* cysts and viruses by free chlorine, chloramine, chlorine dioxide, and ozone. Water utilities use the CT data as a guide to managing chemical disinfection.

The basis for this approach to determining the effectiveness of disinfection technologies is the presumption that laboratory results with test organisms are indicative of results in actual water treatment plants under similar conditions of pH, temperature, disinfectant type and residual, and contact time. The approach appears to have developed in part because of the high level of skill needed for the testing, making universities and research laboratories the appropriate settings for carrying out studies, and in part because of the way EPA has approached disinfection regulation and management. The CT concept, while oversimplifying disinfection kinetics, offers water treatment plant operators a practical way of assessing the adequacy of their disinfection practices, and this is a key factor in managing disinfection in treatment plants.

While the *CT* concept provides a means for evaluating the effectiveness of chemical disinfectants, no national regulatory guidance is available for ultraviolet disinfection of water by public water systems. UV systems would typically be used only by small systems with ground water sources, and ground water disinfection is not regulated by the EPA (as of June 1996).

Corrosion Control Systems

Current regulations allow small systems to install corrosion control systems without pilot tests. The Lead and Copper Rule allows small systems (but not large ones) to select corrosion control strategies based on desk-top reviews of documents and records of water quality because the cost for long-term corrosion control pilot studies would likely be prohibitive for small systems. Under this rule, small systems were to begin monitoring tap water for corrosion-related problems in July 1993; systems requiring corrosion control are to have treatment installed by January 1998.

One alternative to performing a corrosion control study for a small system is to have a state drinking water regulator or other knowledgeable authority review the quality of the water involved and recommend pH or alkalinity adjustments, use of a corrosion-inhibiting chemical, or a combination of these strategies. Another is to implement a corrosion control strategy in use at another nearby system if both use the same source water and treat it in similar ways. The latter option is especially appropriate for ground waters coming from the same aquifer.

MATCHING OPERATOR SKILL TO EQUIPMENT COMPLEXITY

If any package treatment equipment or treatment process requires skilled operation in order to work effectively, certification of the equipment and approvals for its use should incorporate provisions for proper operation. The skill level of the operators needs to be commensurate with the skill level requirements imposed by the equipment being used. Small systems should never accept the contention that "this equipment runs itself and you don't need an operator." State regulatory agencies, consulting engineers, and equipment manufacturers need to discuss this issue and find alternative approaches to ensuring the level of operation needed for the successful application of treatment technology; Chapter 6 recommends ways to improve training of small system operators.

The skill level or type of understanding essential for successful operation of process equipment varies from process to process. Lime softening plants and those incorporating coagulation require some knowledge of the chemistry associated with the processes. Generation of ozone often involves not only the actual ozone generation step but also an air preparation step. A considerable amount of mechanical and electrical equipment can thus be involved, as contrasted to a simple chemical feed pump if sodium hypochlorite is used as a disinfectant. If process equipment manufacturers make extensive use of sensors and automated analytical techniques, such as streaming current detectors, turbidimeters, pH sensors, and so forth, the small system using such technology will either need to have an operator who understands electronics and instrumentation and can keep everything in good repair, or it will need to have rapid-response service contracts so that the instrumentation can be repaired quickly if it malfunctions.

CENTRALIZING DATA COLLECTION

Currently, there is no one centralized data base or clearinghouse for information on the performance of drinking water treatment technologies. Several organizations have data bases or other sources of information on treatment technologies for small systems (see Box 4-2), but the data bases cover only a limited number of technologies, lack standard data reporting formats, and often are missing information on the full range of parameters (for example, raw water quality, finished water quality, and operation and maintenance costs) necessary to evaluate technology performance.

One result of the lack of central data collection is that for all but conventional technologies that have a long record of data, considerable "reinvention of the wheel" occurs in testing. A second result is that, lacking accurate, current information, many engineers, utility managers, and local decision makers continue to select the best-known technology for their system, even if it is not the best choice. Information on alternative technologies and package plants must be made available to these decision makers in an organized and prompt fashion.

The EPA should establish a national clearinghouse to serve as a repository for data obtained on the operational efficiency of treatment technologies. Dissemination of available information on treatment technologies may assist in their acceptance and reduce the expense involved in their adoption. This is especially true for innovative or alternative treatment technologies and systems sold as package plants. Information in the clearinghouse should be made available to all interested parties, including state regulators, water utility managers, consulting engineers, and equipment manufacturers and suppliers.

This central clearinghouse could be established by expanding the Registry of Equipment Suppliers of Treatment Technologies for Small Systems (RESULTS) data base at the National Drinking Water Clearinghouse (NDWC) (see Box 4-2). RESULTS could be expanded to include information on both raw and finished water quality, operational requirements, operation and maintenance costs, and useful life of the technology. Manufacturers could provide lists of installed equipment, and the performance of the equipment at each location could be included. Pilot-scale data, in addition to data from full-scale operations, could be entered into the data base. Periodically, the NDWC or another organization identified by the EPA could evaluate the data to assess the performance limits of different technologies.

For the information in the clearinghouse to be useful in comparing technologies, it must be reported in a standard format. Therefore, a standard testing protocol should be developed, and any test data entered into the data base should follow the protocol.

State agencies responsible for regulating drinking water systems should assign an individual to serve as a liaison to the central clearinghouse. This individual would be responsible for staying informed of performance data for the systems of importance in his or her state. The states could then continually update their requirements for site-specific testing of water treatment systems. For many systems, testing requirements can be decreased as more data become available on system performance under a range of water quality conditions.

CONCLUSIONS

Small systems can expend significant sums in pilot testing water treatment technologies prior to installation. While site-specific testing requirements cannot be entirely eliminated, they can be streamlined. In developing programs to reduce pilot testing requirements, regulators at the state and federal levels will need to consider the following issues:

• **Failure to share water treatment performance data from state to state leads to site-specific pilot testing requirements for package plants that are in some cases unnecessary.** Pilot data collected in one state may not be accepted in another state. In some cases, small water systems must spend money to prove elements of technology performance that have already been demonstrated elsewhere.

• **A nationally accepted program of treatment technology testing and verification could help reduce repetitive pilot testing requirements.** Technology certification would provide assurance to state regulators that a product will perform as advertised and that manufacturers' data are not just a "sales pitch."

• **Even with a national water treatment technology certification program in place, some site-specific testing will be needed before a treatment system is designed and installed.** The testing can be as simple as laboratory tests of water quality parameters or as complex and expensive as multiseason pilot plant investigations. The extent of testing required depends on the type of technology under consideration and the quality of the source water.

• **Surface water treatment technologies for which performance is linked to source water quality generally have more complex testing requirements than technologies whose performance is largely independent of water qual-**

**BOX 4-2 Sources of Information on
Small System Technologies**

The following sources provide a starting point for obtaining information about technologies for small drinking water systems. In addition to these organizations, a regional phone book may have a listing of engineering consulting firms that specialize in water treatment plant design and firms that provide certified treatment plant operation and maintenance services on a contract basis.

• *EPA Safe Drinking Water Hotline (800-426-4791):* The Safe Drinking Water Hotline provides, among other information, technical publications on drinking water. Among the (free) items available from the hotline are an exhaustive list of drinking water publications, of which many are directed to the small water system, published by the agency's Office of Ground Water and Drinking Water (EPA, 1994), as well as a pocket guide to the requirements for operators of small water systems (EPA, 1993). The hotline can also provide contacts for specific information on local drinking water, bottled water, and home water treatment units and names of state contacts who can provide callers with a list of EPA-certified drinking water laboratories in their local area. Hotline representatives will not discuss or recommend specific manufacturers.
• *American Water Works Association (AWWA) Small System Hotline (800-366-0107):* AWWA operates a toll-free "informational-support" hotline, free to water systems in the United States and Canada with fewer than 1,000 service connections. Among other services, the hotline provides information about water treatment and technology options, access to all AWWA resources, information on local resources, and an opportunity to network with others involved in small water systems.
• *National Drinking Water Clearinghouse (NDWC) (800-624-8301):* The NDWC, created by the EPA and the Rural Development Administration, provides a

---

ity. For example, coagulation and filtration systems are more likely to need site-by-site evaluation than membrane filtration systems.

RECOMMENDATIONS

• **The EPA should continue the technology verification program, to be implemented by the National Sanitation Foundation, for water treatment technologies for small systems.** Equipment evaluation should include verification of the manufacturer's claims, especially claims of reductions in contaminant levels, and evaluation of operation and maintenance manuals to ensure that they provide accurate and complete information about the equipment. The verification report should indicate the level of operator oversight required for proper technology performance. After successful testing, technologies should carry a stamp or marking to identify that their performance has been verified.

database of water treatment suppliers for small systems known as the Registry of Equipment Suppliers of Treatment Technologies for Small Systems (RESULTS). RESULTS includes information on treatment effectiveness, contaminants addressed, and suitability of a process for a given source water quality. It provides names and phone numbers of system managers for installed and operating systems of each type from each manufacturer. A small system manager considering a certain treatment technology can then contact and receive information directly from a user of a technology similar to the one under consideration.

NDWC provides RESULTS, which runs on any IBM-compatible computer, for a nominal charge, approximately the price of a computer diskette. RESULTS can sort information by contaminant type, technology, plant location, equipment supplier, or cost. Multiple criteria can be linked to derive information on a system that, for example, removes *Giardia* cysts and costs less than $50,000. RESULTS is limited in that some small system technologies are not included in the data base, information is not reported in a standard format, and important parameters for technology evaluation (such as operating costs and raw or finished water quality) are missing from some of the entries.

The NDWC is housed at West Virginia University along with its "sister" organizations: the National Small Flows Clearinghouse for wastewater treatment technologies and the National Environmental Training Center for Small Communities, a group that provides training resources for drinking water, wastewater, and solid waste treatment. In addition to RESULTS, NDWC offers a quarterly newsletter, a toll-free technical assistance line, a toll-free electronic bulletin board called the Drinking Water Information Exchange, and many free or low-cost educational products.

• *State drinking water agencies:* State agencies that handle drinking water regulatory issues may have information on successful (or unsuccessful) application of certain technologies on waters similar to a given source water.

• **The EPA should establish a standardized national data base for water treatment technology information by expanding the existing RESULTS data base at the National Drinking Water Clearinghouse.** All data base entries should include quality assurance information. The EPA should permit anonymous data entries to allow those providing data to include all reliable data, not just data that comply with regulations. As with the current RESULTS data base, the information should be made available at a nominal fee and should be configured for use on a desk-top computer. It should also be made available electronically, via the Internet. The availability of the data base should be advertised to regulators, water utilities, and consulting engineers.

• **The EPA should oversee the development of standard protocols for pilot testing of water treatment technologies.** These protocols will have to be developed for each treatment technology separately but should allow information gained from pilot tests to be entered into the standardized data base and shared with other potential technology users. Pilot plant data should be made available in the national data base to allow rapid dissemination of this information for use by utility decision makers and state regulatory agencies.

• **The language of certification and testing should be standardized and should be the language of the Safe Drinking Water Act regulations.** Data on raw and finished water quality should be collected in common units corresponding to the requirements of the SDWA.

• **State regulatory agencies responsible for overseeing drinking water systems should establish a mechanism for reviewing and updating their requirements for the testing required before a drinking water plant can be installed.** As more experience is developed on treating types of water important in the state, the amount of site-specific testing required can be decreased.

REFERENCES

APHA, AWWA, and WEF (American Public Health Association, American Water Works Association, and Water Environment Federation). 1992. Standard Methods for the Examination of Water and Wastewater, Eighteenth Edition. Washington, D.C.: APHA.

Crittenden, J. C., P. S. Reddy, H. Arora, J. Trynoski, D. W. Hand, D. L. Perram, and R. S. Summers. 1991. Predicting GAC performance with rapid small-scale column tests. Journal of the American Water Works Association 83(1):77–87.

EPA (Environmental Protection Agency). 1993. The Safe Drinking Water Act: A Pocket Guide to the Requirements for Operators of Small Water Systems. San Francisco: EPA, Region 9.

EPA. 1994. Office of Ground Water and Drinking Water Publications. EPA 810-B-94-001. Washington, D.C.: EPA, Office of Ground Water and Drinking Water.

GAO (U.S. General Accounting Office). 1994. Drinking Water: Stronger Efforts Essential for Small Communities to Comply with Standards. Washington, D.C.: GAO.

Kavanaugh, M. C., and R. R. Trussell. 1980. Design of aeration towers to strip volatile contaminants from drinking water. Journal of the American Water Works Association 72(12):684–692.

Kostelecky, J. D., M. C. Ellersick, W. W. Trask, Jr., B. M. Coffey, and D. A. Foust. 1995. Implementation of microfiltration for metropolitan's small domestic water systems. Presented at 1995 AWWA Membrane Technology Conference, Reno, Nevada, August 13–16, 1995.

Leland, D. D., and G. S. Logsdon. 1991. Pilot plants for slow sand filters. Pp. 191–227 in Slow Sand Filtration, G. S. Logsdon, ed. New York: American Society of Civil Engineers.

McCarthy, R. 1995. Presentation to the National Research Council's Committee on Small Water Supply Systems, Washington, D.C., March 2, 1995.

McClelland, N. I. 1994. NSF International: programs and services offered internationally. Presented at U.S. Russia Business Development Committee Standards Working Group, Moscow, Russia, May 24–25, 1994.

Oxenford, J. L., and B. W. Lykins, Jr. 1991. Conference summary: practical aspects of the design and use of GAC. Journal of the American Water Works Association 83(1):58–64.

Syrotynski, S., and D. Stone. 1975. Microscreening and diatomite filtration. Journal of the American Water Works Association 67(10):545–548.

Taylor, J. S., and L. A. Mulford. 1995. Membrane protocol to meet the ICR. In Proceedings of the 1995 Membrane Technology Conference. Denver: American Water Works Association.

WMA, Inc. 1994. Small Systems BAT Task Force: Interviews with State Officials Regarding the Application of the Recommended Standards for Water Works in Reviewing Small System Technologies. Report prepared for the EPA. Alexandria, Va.: WMA, Inc.

5

Ensuring Small Water Supply
System Sustainability

Since passage of the Safe Drinking Water Act (SDWA), debates among regulators, policymakers, the water supply industry, and other interest groups about how water systems should balance health and safety requirements against the need to contain costs have generated an incremental, issue-by-issue approach to managing water systems (Cromwell, 1994a). A more comprehensive approach is needed. Rather than focusing solely on how to comply with the latest regulations, water systems and regulators need to assess the sustainability of these systems—that is, their long-term ability to provide adequate water service while adapting to new regulations and customer demands.

This chapter discusses processes for evaluating the sustainability of small water systems. It also examines options for improving management of water systems that face challenges in maintaining sustainability.

WHAT IS SUSTAINABLE?

A sustainable water system is one that can meet performance requirements over the long term. Such systems have the following characteristics (Wade Miller Associates, 1991; Okun, 1995):

- a commitment to meet service expectations;
- access to water supplies of sufficient quality and quantity to satisfy future demand;
- a distribution and treatment system that meets customer expectations and regulatory requirements; and

- the technical, institutional, and financial capacity to satisfy public health and safety requirements on a long-term basis.

Like any good business, a sustainable water system can also adapt to future changes in regulatory requirements and customer demand.

Sustainability depends not only on a system's capacity and capabilities, or on its financial prospects, but also on the larger socioeconomic and resource environment that both supports and draws on the system, the regulatory requirements the system must meet, and the technical and financial assistance available to it. A thorough evaluation of all these factors (including the system's own resources) is necessary to identify the main deficiencies jeopardizing sustainability. Water systems need to periodically evaluate their present and future plans, processes, skills, and services and seek to identify options and solutions that will promote sustainability. The ability to effectively implement such potential solutions, and to identify and implement them in the future, is an oft-overlooked but critical component of a system's sustainability. Achieving sustainability is not a one-time task; it requires a continuous effort.

System Capacity and Capabilities

Small systems today face severe challenges, including rapidly increasing regulation, declining water quality and quantity, legal liability for failing to meet the SDWA or other purveyor responsibilities, financial distress, and customer resistance. A system's ability to deal with these challenges depends to a great degree on its managerial, technical, and financial capabilities. Many systems possess adequate staff, expertise, and other resources to meet these challenges, or they can develop the necessary resources. Systems that lack these assets or the ability to develop them, or that simply face community, socioeconomic, or environmental issues beyond their control, usually need to be *restructured*—that is, absorbed into, combined with, or served by other utilities.

Socioeconomic and Environmental Factors

Communities require water of sufficient quality and quantity to meet their needs, reliably delivered at affordable rates. A small water system's ability to meet these basic expectations often depends on factors beyond the system's control.

The most obvious such factor is a lack of sufficient community income to maintain the water system's infrastructure and operations. As discussed in Chapter 2, many small rural systems suffer from this problem.

Another factor is the stability of the community's population. If a system's population is continually decreasing, as in many rural communities, or rapidly increasing, as in many periurban areas, the system often lacks the expertise or resources to deal with the way these changes affect revenues, infrastructure, and staffing responsibilities. Small rural communities in particular frequently lack adequate personnel or capital to manage a public water supply without some form of assistance or restructuring.

Another growing problem is the availability of water resources. This issue now stretches beyond technical or logistical concerns and into social, political, and environmental concerns; what was once the province of engineers and well drillers is now an arena for municipal officials, various coalitions, and differing interest groups. If recent conflicts over the allocation of waters in the Pacific Northwest, in which concerns over maintaining stream flows sufficient to support fish populations are making allocation decisions difficult, and the reallocation of water from the City of San Antonio, Texas, in which concerns over withdrawals from a major aquifer containing endangered species complicated allocation decisions, are any indication, such supply issues will pose a growing challenge to the sustainability of many water systems. The "total water management" concept promoted by the American Water Works Association, which emphasizes the need to account for all uses and sources in water management decisions, simply formalizes the growing recognition that water is a limited resource subject to competing uses. Policymakers need to develop innovative water resource allocation solutions to promote sustainable water service—and small systems must share the responsibility of balancing competing water resource allocation needs.

Community or other political resistance to change is another challenge frequently facing systems trying to address supply or treatment problems. A small system seeking to solve a supply problem by tapping a new source, for instance, may meet resistance either within the community or from interest groups outside of the community because of fears of environmental or other impacts of developing or using the new source.

Similarly, in cases where the solution to a system's problems lies in relinquishing control to another water authority or entity, the decision to restructure may meet resistance from community members, developers, or local officials. In some cases, one community criterion for a successful solution may be that the system function within existing political or community oversight structures.

REGULATORY OPTIONS FOR PROMOTING SUSTAINABILITY

Regulators can employ a wide range of options for regulating small water systems; these options will be best used if all states develop strategies that seek not only to ensure regulatory compliance but also to encourage system sustainability. Options for ensuring sustainability range on a continuum from policies of nonproliferation, which discourage the creation of new small systems, to policies of assistance and support, which bolster existing systems. These approaches are frequently mixed to some degree. For example, a state could

---

BOX 5-1 State and Regional Planning

Several states have taken steps to mandate or encourage regional or state-wide water system planning efforts.

Maryland, through the Maryland Department of Environment (MDE), requires that counties develop comprehensive water supply plans that specify service areas' projected needs for new service over the next 10 years and how any proposed new water systems will be financed. These regulations also call on county authorities to develop specific planning regulations and requirements. The water supply plans must be submitted to MDE for approval and must be updated every 2 years. MDE has the authority to require connection to public water systems or require designs that will facilitate future interconnection to public water systems.

In the Washington and Connecticut programs, final approval authority for regional water supply planning rests with the state. Both states use this combined planning process to assign local officials the responsibility of guaranteeing the service responsibilities of new small systems. In Washington, the establishment of a strong state planning effort required continued efforts directed at the legislature over a period of years. In Connecticut, a severe drought provided the impetus to implement such an effort.

Recognizing that a financial incentive may be required to get agreement at the local level, Washington State provides grant monies in a "matching" format to promote comprehensive assessment or planning efforts. Washington has also developed a financial viability test that new systems must pass before being allowed to be formed.

Pennsylvania has adopted an incentive-based approach. Three demonstration programs have been implemented. One offers regionalization feasibility planning grants to any group of two or more municipalities in rural areas. Another provides demonstration grant funding to study the feasibility of establishing county-wide authorities. The third provides demonstration grants to counties interested in launching comprehensive water supply planning initiatives. This voluntary approach to initiating comprehensive assessments and planning will probably leave some parts of the state uninvolved in this kind of planning effort, but it will encourage planning in many others (Cromwell, 1994b).

---

establish regulatory criteria that identify and discourage unsustainable small systems but that lend assistance to those that show they are or can be sustainable.

A few states have begun to use performance assessments to evaluate water system sustainability. These assessment efforts have focused primarily on regional and local planning policies. For example, Connecticut, Maryland, and Washington have created comprehensive regional and system planning programs through legislation and administrative rules. These are described briefly in Box 5-1. Box 5-2 outlines in more detail the key components of the Washington State regional planning process, which includes both large and small systems.

Such regional or statewide planning efforts provide an effective, economical way to identify problems that should be addressed on a regional basis. They also

BOX 5-2 Elements of the Washington State Regional Water System Planning Process

I. Preliminary assessment
  A. Existing water systems
    1. History of water quality, reliability, service
    2. Fire fighting capability
    3. Evaluation of facilities
  B. Future water sources
    1. Availability
    2. Adequacy
  C. Service area boundaries
    1. Map of established boundaries
    2. Identification of systems without boundaries
  D. Growth in the area
    1. Current population and land use patterns
    2. Population and land use trends
  E Status of planning
    1. Water system
    2. Land use
    3. Coordination

II. Individual water system plans
  A. Basic planning data
    1. Service area description
    2. History of system (planning, sources, etc.)
    3. Present and future land use
    4. Present and future population
    5. Present and future water use
  B. Inventory of existing facilities
    1. Description of existing sources and system facilities
    2. Hydraulic analysis
    3. Water quality and conformance with standards
    4. Fire fighting capability

  C. System improvements
    1. Projection of 10-year water demand
    2. Description of alternatives to meet demand (and their costs)
    3. Selection and justification of alternative
    4. Schedule of improvements
    5. Financial program
  D. Other topics
    1. Watershed control program
    2. Service area agreements
    3. Analysis of shared facilities (interties, reservoirs)
    4. Relation between water and land use plans
    5. Operations program
    6. Consideration of state Environmental Policy Act
    7. Maps supporting the plan

III. Area-wide supplement
  A. Assessment of related plans and policies
  B. Future service areas in the region
  C. Minimum areawide design standards
  D. Process for authorizing new water systems
  E. Future areawide source plans
  F. Plans for development of joint use or regional facilities
  G. Application of satellite support systems
  H. Other topics pertaining to the region
  I. Compatibility of supplement with other plans and policies
  J. Continuing role of Water Utility Coordinating Committee
  K. Consideration of state Environmental Policy Act

provide opportunities to assess and coordinate alternatives that promote sustainability at both the regional and individual system levels. Regional planning processes, for instance, can identify systems that have difficulty conducting an assessment effort and thus may be prone to failure. This early identification of troubled systems is important to prevent "throwing good money after bad" by trying to maintain a system doomed to failure. Experience demonstrates that once individual systems begin to spend money to keep service afloat, the system administrators or owners often give increasing resistance to relinquishing control of the system, even when doing so is the most cost-effective alternative (Cromwell, 1994).

This situation is aggravated by the phased implementation of many SDWA regulations. A larger system that might lead a restructuring effort through direct connection or development of a satellite strategy, for instance, may face compliance requirements several years before neighboring small systems will, creating a situation in which the smaller systems have no motivation to cooperate or join with the larger system's effort. Also, depending on its source of supply, a water system may have to comply with certain rules years before a neighboring system with a different supply source does, leaving the second system disinclined to cooperate with the first in any collaborative efforts. Thus, while these varying compliance deadlines may allow sustainable small systems time to make necessary changes, they can also leave serious problems in unsustainable systems hidden if those systems go unevaluated. A coordinated regional planning process will help reveal such problems early so that the troubled systems can either be fixed or merged with other systems.

PUBLIC HEALTH PERFORMANCE APPRAISALS

Whatever a state sees as its regulatory role, its central motivation should be to protect public health and ensure reliable water service at a reasonable cost. This requires evaluating the performance of small water systems in a structured and objective manner. Mandatory, comprehensive public health performance appraisals can serve a critical role in this evaluation. Performance appraisals can

- assist regulators in evaluating SDWA compliance and the sustainability of systems;
- allow a system to comprehensively review its strengths, deficiencies, and needed improvements;
- present a rationale for specific internal management improvements or restructuring options;
- ensure that consumer expectations of being provided reliable, sustainable, and high-quality water service are met.

Perhaps most important, a requirement for periodic performance appraisals enables state regulatory agencies to be proactive in ensuring safe water and promoting sustainability, rather than using piecemeal reactive measures. The U.S. Environmental Protection Agency (EPA) should award state revolving fund (SRF) monies for drinking water systems only to states with structured programs for conducting public health performance appraisals of water systems regulated by the SDWA. This will ensure that SRF monies are not used to prop up unsustainable small systems and that the funds are used as cost effectively as possible.

The basis for the public health performance appraisals could be an evaluation form developed by state regulators and sent to all community water systems. For very small systems, the evaluation form need not be more than two pages long. Initially, the performance appraisals could be phased in over a 5-year period to allow a manageable work load for state regulators. When a water utility's performance appraisal reveals inadequacies in the system, states will need to provide assistance in developing options (such as management changes and restructuring) to correct the inadequacies, so that water service to the community can be maintained.

Figure 5-1 shows suggested key components of a model state performance appraisal program. To enforce the performance appraisal requirement, states can impose regulatory sanctions such as the denial of a proposed system expansion or the rejection of state funding applications for systems that fail to comply. States can also condition the receipt of an operating permit on successful completion of a performance appraisal, as shown in Figure 5-1. This creates strong incentives for compliance through its effect on the banking and mortgage industry, since many lenders will approve residential or commercial purchase, refinance, or construction loans only if the applicants are hooked up to permitted water systems. In Washington State, for instance, any public water system that must meet federal requirements, as well as any satellite system management agency, must obtain an annual operating permit from the state Department of Health for each system owned; to get a permit, the system must conduct a performance assessment. There is an operating permit fee based upon system size and type. The operating permit must be renewed annually or with a change in ownership. Numerous criteria regarding system integrity, reliability, SDWA compliance, management, financial viability, and planning are used to evaluate systems and place them into one of three permit categories: substantial compliance, conditional compliance, or substantial noncompliance. The Department of Health may impose conditions on a permit or modify, revoke, or alter it at any time as changes occur in the system. (Of course, where permits are revoked for existing systems, the state will need to ensure that the affected community continues to receive water service.)

Several other states have used performance appraisals or similar programs to support policies of nonproliferation. These states include Massachusetts, where the Department of Environmental Protection may deny system approval unless the system demonstrates that it has the technical, managerial, and financial re-

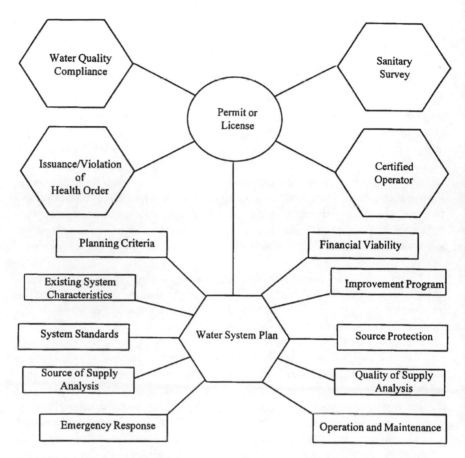

FIGURE 5-1 Model state public health performance appraisal program, in which a performance appraisal and development of a water system plan are required in order to obtain an operating permit or license.

sources to operate and maintain the system; Montana, where the Department of Health and Environmental Sciences may review the financial viability of new or expanding water systems; and Idaho, where the Idaho Public Utility Commission may deny certifications for proposed new investor-owned utilities if there is no need for the service or if another company is willing and able to provide similar or better service.

Elements of a Public Health Performance Appraisal

The public health performance appraisal process (or licensing or permit criteria process) should require documentation and evaluation of several key indicators of system performance and sustainability. These indicators include the following:

• *Record of issuance of health orders*: Any issuance of a health order or other water-quality-related health violation calls into question a system's ability to reliably deliver safe drinking water. Requiring automatic moratoria on new customers for systems issued health orders will help compel correction of unacceptable conditions.

• *Record of water quality compliance*: Complying with monitoring and maximum contaminant level (MCL) water quality standards is the most fundamental responsibility of any water utility and should be required for licensing or permitting. Monitoring requirements should include specifications for accurate recordkeeping.

• *Certification of operators*: A system's size, treatment process, supply and distribution characteristics, and other operational parameters will dictate the number of staff and their appropriate training or certification level; permitting standards should require such appropriate training or certification, as well as continuing education.

• *Record of sanitary surveys*: Regular sanitary surveys, or on-site visits, have long been an effective means for regulatory agencies to inspect and aid the operation of small water systems. These visits allow regulators to verify facilities and their condition, evaluate source protection measures and operator abilities, and recommend improvements. They should be part of any assessment or certification program.

• *Existence of a water system plan*: Prior to receiving a permit, the water system should have a plan specifying how it will meet current and future performance criteria.

While the importance of health violation records, water quality compliance, operator certification, and sanitary surveys have long been considered by regulators evaluating small water systems, the importance of a comprehensive, forward-looking water system plan is only now becoming recognized among planners and system owners and operators. For that reason, it is worth examining what these plans involve and the role they play in performance appraisals.

Small Water System Plans

Every water utility should create a comprehensive plan that specifies how the utility will affordably meet present and future demands while complying with SDWA and other regulations. More than any other aspect of the performance appraisal, the creation of a water plan compels the utility to examine itself closely and develop a road map for the future.

A utility's plan should be written somewhat like a business plan. The plan should include information on future trends in the service area, population and growth, land use policies, water demands, and other factors on both a short-term and long-term basis spanning 5 to 20 years. In addition, like any good business, water systems should make customer satisfaction a priority in the planning process and should involve customers in developing their plans. The level of detail included in the plan will vary with the size and complexity of the water system. Example outlines for plans for small, relatively simple systems and larger, more complex systems are included in Boxes 5-3 and 5-2, respectively.

Regardless of the format used for the water system plan, the plan should include the following elements:

• *Evaluation of existing system characteristics*: A water system plan should include a description and inventory of system facilities and their general condition, as well as an analysis of their capability to supply sufficient water quality and quantity to meet existing and projected demands. The analysis should include an evaluation of raw and treated water quality for each source and the distribution system and of the condition, capacity, and reliability of the source of supply, storage facilities, and system piping. Finally, this evaluation should discuss possible remedies for any existing and anticipated deficiencies in these areas.

• *Statement of system standards*: The plan should list all federal, state, or other standards the system must meet. These may include water quality parameters, average and maximum daily demands, peak hour demand, storage requirements, fire flow rate and duration, minimum system pressure, minimum pipe sizes, telemetry and back-up power requirements, valve and hydrant spacing, and other policies that affect performance and design. In addition, the utility should establish standard construction specifications for system expansion.

• *Analysis of source of supply*: The plan should evaluate opportunities to optimize the use of existing water sources and evaluate both possible new sources and other innovative methods to meet water needs. This analysis should consider the pros and cons of developing and implementing a conservation program; confirming or changing water rights; pursuing interties with neighboring utilities for either emergency or sustained usage; and artificially recharging available surface water into an aquifer and subsequently withdrawing it.

• *Plans for source protection*: The water plan should include a program to protect or improve source waters through the development of a wellhead protection and/or watershed control program.

• *Statement of operation and maintenance program*: In many small water systems, operations procedures are known only to the system operator. Carefully documenting what is involved in operating and maintaining the system will allow a smooth transfer of this knowledge so that the utility can continue to provide good service if the operator leaves the utility's employment or for some reason cannot run the system. The operation and maintenance program should describe

---

**BOX 5-3 Elements of a Water System Plan for
Very Small Systems**

I. Basic planning data
   A. Service area description and map
   B. History of system (planning, sources, health violations, etc.)
   C. Current and future population
   D. Current and future water use

II. Inventory of existing facilities
   A. Description of existing sources and system facilities
   B. Service performance history (adequate pressure, reliability, etc.)
   C. Water quality and conformance with standards
   D. Fire fighting capability and conformance with standards (if necessary)

III. System improvements
   A. Projection of future water demand
   B. Description of alternatives to meet demand and their costs
   C. Selection and justification of alternative
   D. Schedule of improvements
   E. Financial program

IV. Management program
   A. Structure/organization chart
   B. Staffing requirements
   C. Certification and training required
   D. Operational and emergency procedures

V. Financial program (financial viability test)
   A. Budgets
      1. Operations, maintenance, general and administrative
      2. Replacement and improvements
   B. Reserves
   C. Rates (history)

---

the primary day-to-day operational activities and state by whom and how they are performed. Suggested elements include job descriptions for management and operating personnel, along with a description of the decision-making "chain of command"; descriptions of system components and how they function and are controlled; preventive maintenance activities and their scheduled frequencies; equipment and chemical inventories and suppliers' names, contacts, addresses, and phone numbers; safety procedures; cross-control connection activities; and a listing of the appropriate recordkeeping procedures and reporting requirements.

• *Emergency response program*: The plan should clearly specify appropriate remedial action and notification procedures, call-up lists, and back-up procedures for times that normal operational conditions are disrupted. This response plan should include a realistic vulnerability analysis that identifies the major facilities at risk under a variety of emergency situations and operational responses to their failure in different scenarios.

• *Analysis of quality of supply*: SDWA requirements have become increasingly complex for both source and distribution monitoring and compliance. A plan for monitoring and reporting all water quality testing parameters will help system operators and managers meet SDWA standards. This element of the plan might address monitoring locations for each parameter; sampling schedules; sampling waivers and conditional requirements; lists of certified testing laboratories; any special monitoring requirements; and procedures to follow in the event of an MCL or sampling violation.

• *Statement of improvement program*: This vital element of the plan specifies the capital and operational improvements needed to rectify problems the plan has identified. The improvements should be prioritized and scheduled and their costs estimated. Prioritization of these improvements will be influenced by the health risks, project costs, funding options, related improvements, and other policy considerations relevant to each option. This improvement program will be vital to creating a detailed budget of capital and operational expenses and to devising a suitable financing strategy for implementation.

• *Financial viability evaluation*: Finally, every water plan should include a financial viability evaluation that assesses the system's ability to remain financially solvent while complying with all relevant regulations. This is one of the most critical parts of the plan and as such warrants detailed description.

Financial Viability Evaluations

A financial viability evaluation seeks to measure a system's "ability to obtain sufficient funds to develop, construct, operate, maintain and manage a public water system on a continuing basis in full compliance with federal, state, and local requirements." (WADOH, 1994). The evaluation identifies the total cost of providing water service, including needed improvements, and establishes fees adequate to cover these costs. In general, the financial impact of improvements is usually much greater on small systems than on those with a larger customer base. However, various experiences in Pennsylvania, Washington, and elsewhere have shown that it is often possible for small systems to budget and pay for necessary improvements.

The financial viability evaluation takes into account how the system will be affected by such factors as more restrictive funding sources for private versus public utilities; the density of the system's customer base compared to the area served; the system's prospects for consolidating financial resources; and any past practices that have created high costs. A utility may not be able to control all of these factors, but it can recognize or anticipate them and plan to minimize their impacts.

One useful model for such evaluations is the evaluation process developed by Washington State for small systems. The evaluation seeks to determine whether the utility is in compliance with the operating and monitoring requirements imposed by the SDWA in the short term and to determine whether the system is sustainable, from a financial resource perspective, over the long term. The evaluation includes four key "tests" that assess whether a water system can operate successfully day to day, respond to emergency situations, plan for needed improvements and expected growth, and develop and maintain a reliable base of financial support.

These tests include the following:

- *Test 1: Is there a workable operating budget?* The operating budget test reviews whether a system is generating sufficient revenue to meet its estimated expenses. If the system lacks sufficient revenue to meet expenses, it should either raise its water rates or reduce nonessential expenses.
- *Test 2: Is there an adequate operating cash reserve?* This test evaluates the system's ability to withstand cash-flow fluctuations. There is often a significant length of time between when a system provides a service and when a customer may pay for that service; a study of the system's historic cash flow can accurately quantify the time period between delivery and payment for service. Since a 45-day difference is the industry norm, most systems attempt to keep at least 45 days' worth, or one-eighth, of their annual operating and maintenance and general and administrative expenses in an operating cash reserve to prevent potential cash-flow problems. The operating cash reserve is essentially the "checkbook balance" a system should maintain to meet cash-flow needs and pay for unforeseen operating emergencies. If a system does not presently have an existing operating cash reserve equal to or greater than one-eighth its annual operating budget, it needs to develop one with (1) a one-time charge, (2) a transfer of funds from an existing reserve, or (3) funds accumulated over a reasonable period of time through the budget process.
- *Test 3: Is there an emergency reserve fund?* This evaluates the system's ability to cover the costs of the failure or loss of its most vulnerable system component (generally a production well, a source of supply, the largest pumping equipment, or key transmission lines). Such a reserve can be funded initially with a one-time charge, a transfer of funds from existing reserves, funds accumulated over a reasonable period of time through the budget process, or an alternative financing arrangement.
- *Test 4: Will water rates be affordable based on local criteria?* In this fourth and final test, the system measures the rate impact that any expected increased operating and facility expenses will have on its users and provides an indication of a residential connection's ability to pay the existing and projected

rates. To conduct this test, a system can determine the current and projected average annual residential water bill and compare it to the local average annual median household income, or MHHI, which is a value computed by the U.S. Census Bureau. If the expected rates exceed a certain percentage of the MHHI for the county, the system's costs are too high. The appropriate percentage of the MHHI used as a ceiling for water rates varies by location depending on factors such as the availability of water resources and citizens' willingness to pay for water. Washington State regulators use 1.5 percent of MHHI as a rule of thumb for determining whether water rates are reasonable, but, especially in arid climates, citizens may be willing to pay more. In Washington State, when water rates exceed 1.5 percent of the MHHI, the utility must attempt to identify a more cost-effective means of providing service.

It is presumed that a water system can control the outcome of the first three tests (i.e., the water system either does or does not perform the required actions). The fourth test, however, is only to be used as a tool in determining whether the rates are affordable. It may not be in the power of the water system to ensure that its water rates are less than 1.5 percent of the MHHI.

In Washington State, failing these tests can result not only in denial of water system permits and the associated denial of building permits by local governments, or of home mortgages by lending institutions, but also in the promotion of active restructuring efforts.

OTHER NONPROLIFERATION TOOLS

Public health performance appraisals are just one tool that states can use to curb the proliferation of small water systems. The various tests used in performance appraisals can be used separately, and in a variety of ways, in nonproliferation policies. Such policies often require interagency coordination to provide effective regulation and assistance, since water system approval and oversight seldom fall under the purview of a single entity. In Connecticut, for instance, the drinking water agency and the public utility commission jointly issue operating permits and share the tasks of reviewing site, source, construction, and financial requirements. In that state, local authorities that grant local permits in the absence of a state operating permit may ultimately be held responsible for providing water service if the system in question fails. Maryland, with a relatively strong tradition of local planning, requires that permit approval requirements be specified in local subdivision and land development regulations.

A number of states condition water system approval on special requirements designed to ensure sound management and financial practices. The Maryland Department of the Environment can require escrow accounts, sinking funds for replacement (a type of reserve fund), and performance bonds. The department may waive these financial requirements in cases where the water system has a binding public works agreement with the county government; such agreements stipulate the terms under which ownership will be transferred to the local government entity.

The Ohio utility commission requires unobligated paid-in capital equal to 40 percent of the construction cost of new facilities, as well as commitments from financial institutions for the remaining 60 percent. Nevada grants approval to privately owned small water systems where no alternative system is available, but system operators must post a 5-year performance bond with the local governing body, which has the ultimate responsibility for water service if the system fails. In California, the Public Utilities Commission may require a proposed new investor-owned utility to post a bond of up to $50,000 if gross operating revenues are projected to be less than $200,000. Further, in a survey of state drinking water officials for the American Water Works Association (McCall, 1986), 15 states indicated that their permit review processes require small systems to review and evaluate regionalization, consolidation, contract service, or other alternatives.

For a number of reasons, it is difficult to say at this point how well various nonproliferation efforts are working. However, the number of drinking water systems in this country has grown dramatically, suggesting that proliferation of nonsustainable systems is likely to be a continuing problem, increasing the need for mandatory public health performance appraisals.

UTILITY OPTIONS FOR IMPROVING SUSTAINABILITY

Once a performance appraisal or other method has evaluated how well a water system measures up to public health and safety standards, the utility must work to achieve compliance with regulatory requirements and service expectations either on its own or, if it cannot manage the necessary changes, by restructuring, which involves either entering into mergers or other cooperative arrangements with other (usually larger) systems or transferring management and/or ownership to another entity.

Internal Improvements

While they struggle to meet daily operational and regulatory requirements, small water systems must find ways to make the capital improvements or service enhancements necessary to ensure long-term sustainability. Maintaining this long-term focus in the face of pressing immediate needs is one of the greatest challenges small water systems face.

As is often the case, money lies at the heart of this challenge. Small systems in particular are hampered by limited access to capital markets (a problem both because the relatively small amounts small systems seek to borrow—generally well under $1 million—do not attract institutional investors and because these

TABLE 5-1 Federal Funding Programs for Small Public Water Systems

Contact	Telephone
Appalachian Regional Commission (ARC)	(202) 673-7874
Department of Housing and Urban Development (HUD)	(202) 708-2690
Community Development Block Grants, Economic Development Administration (EDA)	(202) 482-5113
Indian Health Service (IHS)	(301) 443-1083
Rural Utility Services (RUS) (formerly Farmer's Home Administration)	(202) 720-9589

SOURCE: Reprinted, with permission, from Campbell (1993). ©1993 by the *Journal of the American Water Works Association.*

systems often have no borrowing track record); by an insufficient rate and/or tax base, either because the number of customers is small or because the population served has a low per-capita income; by inappropriate rate and fee structures; and by limited managerial resources, which are often, in turn, caused by the system's limited financial resources. This situation is complicated by the difficulty of identifying all available funding resources and by heavy competition for those funds. These challenges require system managers to be creative and persistent in developing a financing strategy.

However, a determined, innovative small system can often find financing alternatives and assistance for infrastructure, operations and maintenance, or other funding needs. Some of the most common financing alternatives are described below. Table 5-1 lists federal funding sources for small water system infrastructure and improvement; Table 5-2 lists state funding sources.

Funds for Capital Improvements

Typical funding alternatives for making improvements include state and federal grant and loan programs, conventional commercial loans (both short- and long-term), and long-term debt-financing mechanisms such as municipal, general obligation, rate revenue, or assessment bonds. Other options include capital facility charges (or "hook-up fees"), paid by new users as they connect to a system, and developer extension policies, in which a developer either pays an "impact fee" to the utility to finance or directly bears the cost of the infrastructure expansion.

One new funding option is the SRF, which is essentially a self-perpetuating loan fund replenished by previous borrowers. In July 1996, Congress established a federally backed SRF program specifically for improving drinking water systems.

Funds for Operation and Maintenance

For operation and maintenance costs, utilities generally turn first to monthly user charges and/or commodity rates (AWWA, 1991, 1992). At minimum, these charges should cover the cost of operating the system at peak capacity; the commodity costs associated with the total consumption of water over a given period of time, the "customer costs," or costs associated with having customers enter or leave the system, and the cost of supplying water for fire protection. These charges are also sometimes used to retire debt service related to capital improvements.

Some utilities also fund operating and maintenance costs with transfers or subsidies from other governmental departments. However, in the interest of self-sufficiency and stability, utilities should not depend on interdepartmental or interfund subsidies except for the purpose of making health-related improvements.

Finally, more utilities are trying to reduce funding needs by pursuing public-private partnerships. The main types of such partnerships are defined in Table 5-3. One frequent form of public-private partnership is the outsourcing of discrete services, such as consulting, meter reading, facility construction, or even system operation and maintenance. Any of these can help a utility stabilize rate costs or improve service.

Restructuring

Restructuring strategies, unlike internal changes to the water system, often require small systems to relinquish some degree of control. Restructuring frequently involves a change in ownership for the utility and almost always involves assistance from an outside restructuring agent. The role played by the restructuring agent depends on individual system needs and plans for improvement. There are numerous examples of these arrangements in place throughout the country.

The relationship a restructuring agent assumes with a troubled utility can take one of four forms: (1) direct ownership, (2) receivership or regulatory takeover, (3) contract service, or (4) support assistance. A brief explanation of each type of relationship and some relevant policy and procedural issues typically evaluated by the restructuring agent before assuming an assistance role are outlined below.

Direct Ownership

Direct ownership transfers responsibility for ownership and operation from the utility in distress to a restructuring agent. This most commonly involves the consolidation of two or more utilities through a merger, acquisition, or regulatory takeover. Water supply may be provided through a new intertie connection to a

TABLE 5-2 State Financing Mechanisms for Small Water Supply Systems

State	Program	Type	Contact
Alaska	Department of Community and Regulatory Affairs, Rural Development Division	Grants and loans	(907) 465-4890
	Department of Revenue, Alaska, Municipal Bond Bank Authority	Loans	(907) 274-7366
	Community Enterprise Development Corporation of Alaska (with Farmer's Home Administration)	Loans	(907) 465-5000
Alabama	Alabama Underground Storage Tank Program	Loans	(205) 271-7720
	State revolving fund—proceeds of bond sales used for loans	Loans	(205) 271-7720
Arkansas	Community Resource Group–Community Loan Fund	Loans	(501) 756-2900
	Water Resources Development General Obligation Bond Program	Bonds and lease–purchase	(501) 682-1611
	Water Development Fund Program	Loans, grants, joint ventures, deferred loans	(501) 682-1611
	Water, Sewer, and Solid Waste Revolving Fund	State revolving fund	(501) 682-1611
	Water Resources Cost-Share Revolving Fund	Cooperative agreement	(501) 682-1611
	Community Loan Fund	Loans	(501) 797-3783
California	Department of Health Services, Division of Drinking Water and Environmental Management	Loans	(916) 323-6111
Connecticut	Connecticut Works Fund Loan Guarantee Program	Guaranteed loans	(203) 258-7822
	Connecticut Development Authority–Business Assistance Fund	Loans	(203) 258-7822
Florida	Florida State Bond Fund	Bonds	(904) 487-1855
Georgia	Georgia Environmental Facilities Authority	Loans	(404) 656-0938

State	Program	Type	Phone
Iowa	Iowa Department of Natural Resources, Environmental Protection Division	Loans	(515) 242-4837
Idaho	Loans-commercial irrigation	Loans	(208) 334-1369
Illinois	Illinois Rural Bond Bank	Bonds	(217) 524-2663
Indiana	Community Promotion Fund	Grants and bonds	(317) 233-8911
Kentucky	Drinking Water Loan Fund (local government, fund B2)	Loans	(502) 564-2382
	Infrastructure Revolving Loan Fund (fund B)	Loans and grants	(502) 564-2090
	Kentucky Infrastructure Authority–Bonding Bank	Bonds	(502) 564-2090
Louisiana	Louisiana Drinking Water Program (connect fee)	Loans	(504) 568-5101
Maryland	Water Supply Financial Assistance Program	Bond money	(301) 631-3706
Missouri	Missouri Department of Natural Resources	Grants	(314) 751-1599
Montana	Renewable Resource Development Program	Grants and loans	(406) 444-6699
	Water Development Program–Department of Natural Resources and Conservation	Grants	(406) 444-6699
	Treasure State Endowment Program–New State Infrastructure Financial Program	Grants and loans	(406) 444-6699
	State Water Plan Advisory Council–Water Development Program	Grants	(406) 444-6668
	State Special Revenue Account for Water Storage	Special fund	(406) 444-6699
	Department of Natural Resources and Conservation—Resource Development Bureau	Grants and loans	(406) 444-6668
New Hampshire	New Hampshire Community Local Fund	Loans	(613) 224-6669

continued on next page

TABLE 5-2 *Continued*

State	Program	Type	Contact
New Jersey	Water Supply Rehabilitation Loan Program–10-year/20-year hardship	Loans	(609) 292-5550
	Water Supply Interconnection Loan Program (shared peak demand)	Loans	(609) 292-5550
	Water Supply Replacement Program (types A and B, shared peak demand)	Loans	(609) 292-5550
New Mexico	Laboratory fee program	Grants	(505) 827-0152
North Carolina	State Revolving Loan and Grant Program–Division of Environmental Health and Natural Resources–Clean Water Program	Grants and loans	(919) 733-6900
Ohio	Emergency Village Capital Improvement Special Account	Advance loans	(614) 644-2829
	Ohio Water Development Authority	Cooperative agreements and bonds	(614) 644-5822
	Ohio Water Development Authority 2% Hardship Loan Program	Loans	(614) 466-5822
	Ohio Water and Sewer Rotary Commission	Loans	(614) 466-2285
	Community programs	Loans	(614) 469-5400
Oklahoma	Water Emergency Grant Program	Grants	(405) 231-2621
	Community Loan Fund	Loans	(405) 231-2621
	Indian Health Service	Grants	(405) 945-6800
Oregon	Water Development Loan Fund–Oregon Water Resources Department	Loans	(503) 731-4010
	Oregon Bond Bank	Grants and loans	(503) 731-4010
	Safe Drinking Water Funding Program	Loans	(503) 731-4010
	Special Public Works Fund–Oregon Health Development	Loans anc grants	(503) 731-4010
	Small-scale energy loans	Loans	(503) 731-4010
	First Interstate Bank of Oregon Foundation	Grants	(503) 225-2167

State	Program	Type	Phone
Pennsylvania	Pennvest–Water Facilities Loan Funds	Loans	(717) 787-8137
	Pennvest–Site Development Capital Budget Projects	Grants and loans	(717) 787-9035
Rhode Island	Development Loan Program	Loans	(401) 277-2217
	Water Facilities Assistance Grant Program	Grants	(401) 277-2217
South Dakota	Governor's Office of Economic Development–Community Projects	Grants	(605) 773-4216
	Governor's Office of Economic Development–Special Projects	Grants and revolving loans	(605) 773-5651
Tennessee	Tennessee Department of Economic and Community Development	Grants	(615) 741-6201
	Tennessee Division of Construction Grants and Loans	Grants and loans	(615) 741-0638
	Tennessee Association of Utility Districts	Loans	(615) 896-9022
	Tennessee Division of Water Supply Reviews, Water Supply Construction Grants	Grants	(615) 741-6636
Texas	Community Loan Fund	Loans	(512) 458-7542
Utah	Utah Drinking Water Board Loan Program	Loans	(801) 538-6159
	Utah Permanent Community Impact Fund	Loans and grants	(801) 538-8729
	Department of Natural Resources–Cities Water Loan Fund	Loans	(801) 538-7294
	Department of Natural Resources–Revolving Construction	Loans	(801) 538-7294
	Department of Natural Resources–Conservation and Development Fund	Long-term loans	(801) 538-7294
Virginia	Virginia Environmental Endowment–The Virginia Program	Grants	(804) 644-5000

continued on next page

TABLE 5-2 *Continued*

State	Program	Type	Contact
	Virginia Environmental Endowment–Water Quality Grants	Grants	(804) 644-5000
	Industrial Revenue Bond Housing and Community Development	Bonds	(804) 371-7061
	Virginia Housing Partnership Fund's Indoor Plumbing Program	Grants and loans	(804) 371-7100
	Drinking Water Bonds	Bonds	(804) 644-3100
	Rural systems	Grants and loans	(703) 345-1184
	Drinking Water Revolving Loan Fund	Loans	(804) 765-5555
Washington	Washington State Public Works Trust Fund (PWTF)–Department of Community Development	Loans	(206) 493-2893
	Washington State PWTF emergency loans	Loans	(206) 493-2893
	Washington State PWTF general construction matching loans	Loans	(206) 493-2893
	Washington State PWTF capital improvement planning	Loans	(206) 493-2893
	Department of Trade and Economic Development (Community Economic Revitalization Board)	Grants and loans	(206) 464-6282
	Department of Ecology, Centennial Clean Water Fund	Grants and loans	(206) 459-6096
	Department of Ecology, Interim Ref 30	Grants	(206) 459-6096
	Washington Bureau of Reclamation Distribution System Loan Act	Loans	(206) 334-1639
	Conservation Commission Water Quality Research Grant Program	Grants	(206) 459-6141
	Department of Ecology, State Revolving Fund	State revolving fund	(206) 459-6061
	Washington Local Development Matching Fund–Department of Community Development	Loans	(206) 586-0662
	Northwest Area Foundation	Grants	(612) 224-9635
Wyoming	State Farm Loan Board	Loans	(307) 777-7781

SOURCE: Adapted, with permission, from Campbell (1993). ©1993 by the *Journal of the American Water Works Association.*

different system or, if water supply is acceptable, through the operation of the troubled system as a satellite operation by the other utility.

Policy and procedural issues that must be addressed in direct ownership takeovers include system size, infrastructure improvement needs, capital improvement needs, purchasing costs, rate structures, and, finally, the system's monetary value, if any. Each of these bears some elaboration.

- *System size*: Some utilities accommodate requests for assistance from systems of any size. Others, however, because of differences in regulatory, rate base, and financial capabilities between large and small systems, might have preferences for systems above or below a certain size.

- *Infrastructure condition*: Some utilities will not accept systems that do not meet their design and construction standards or some other level of minimum qualifications. For systems not constructed in accordance with minimum standards, an engineering evaluation may be necessary to evaluate system upgrade requirements. The costs for this engineering evaluation may be paid or shared by either party.

- *Capital improvement, purchasing costs, and rate structures*: Financing of local improvements and purchasing costs, if any, must either be spread throughout all of the new utility customers or assigned specifically to the residents of the restructured system. This may affect whether identical rates are used for all customers, or whether the restructured customers will pay a capital surcharge in their rates.

- *System value*: Utilities vary widely in how they participate financially in the takeover of small systems. Utilities frequently will not assist if a small system in trouble is seeking a purchase amount. More often than not, the small system is glad to merely transfer the responsibility to the utility at no cost. If money is to change hands, however, a financial feasibility analysis is usually conducted to determine the assessed value of the system or to use as a basis for negotiating the purchase. Any of several methods may be used to establish a system price. *Comparable sales*, where such may be found, may become the basis for the system value. Another method is to use the *net book value*, in which the original installed cost is reduced by accumulated depreciation and contributed capital. *Replacement value* methods estimate the value as if the system were built at current construction costs (but do not account for the cost of needed improvements—a factor that can lead this method to overvalue a system). The *depreciated replacement value* method calculates a value by depreciating the replacement costs of infrastructure from the date of their installation to present. Again, if the system is inadequate and does not meet prescribed standards, then the value of the system, even using this method, will be overstated unless an adjustment is made for the system's functional depreciation. In a *capitalized income* calculation, the annual net income and a rate of return are used to calculate a lump sum

TABLE 5-3 Types of Public-Private Partnerships

Type	Definition	Responsibilities
Contract services	Private partner contracted to provide specific municipal service	Financing: public Design: public-private Construction: public-private Ownership: public O&M^a: private
Turnkey	Private partner designs, constructs, and operates an environmental facility owned by public sector	Financing: public Design: private Construction: private Ownership: public O&M: private
Developer financing	Private party (usually developer) finances construction or expansion of environmental facility in return for right to build houses, stores, or industrial facilities	Financing: private Design: either Construction: either Ownership: either O&M: either
Privatization	Private party owns, builds, and operates a facility and partially or totally finances the operation	Financing: private Design: private Construction: private Ownership: private O&M: private
Merchant facilities	Private company makes a business decision to provide an environmental service in anticipation of profit	Financing: private Design: private Construction: private Ownership: private O&M: private

^{a}Operation and maintenance.

payment equivalent to the present value of the income stream generated by the utility. Finally, a *negotiated sale* may use any combination of the above methods and adjustments as a basis for negotiating a price agreeable to both buyer and seller.

Receivership or Regulatory Takeover

Some states have the authority to take over or transfer management of failing water systems that put public health and safety at risk (see Box 5-4). These transfers sometimes include a transfer of ownership as well. For example, Washington State may place a failing water system under the responsibility of a county

---

BOX 5-4 Receivership: Greenacres Water Supply

Although ordered by the Connecticut Department of Health Services (DOHS) to make a variety of improvements, the owners of Greenacres Water Supply determined that they could not afford the $191,000 required to upgrade their 115-connection system. Instead, they notified DOHS that they wanted to quit the water business altogether.

DOHS asked the state Department of Utility Control (DPUC) to hold a hearing on the matter. During the hearing, two water systems expressed interest in purchasing Greenacres Water Supply and operating it as a satellite system. Later, Greenacres' owners agreed to sell the system to the Tyler Lake Water Company for $10,000, but the DPUC consumer counsel opposed the price as excessive. After examining Greenacres' financial records and considering the improvements that the DOHS proposed, DPUC determined (1) that $617 was a more reasonable price and (2) that the Bridgeport Hydraulic Company (BHC) was a "more suitable entity" to own and operate Greenacres. (BHC already operated the North Canaan water system, and its water mains ran within 4,000 ft of Greenacres Water Supply.)

Ownership of Greenacres was transferred in 1988. The drinking water system that BHC purchased had three wells, one spring, a 6,300-gal and a 2,500-gal atmospheric water tank, and a 5,000-gal pressure tank. The distribution system consisted of 11,583 ft of 1- to 2-in. galvanized, plastic, and copper pipe. None of the 107 residential, 1 commercial, or 7 industrial customers was metered. There was no fire protection.

The state ordered BHC to spread the cost of system improvements across its base of 96,000 customers to reduce the financial burden on Greenacres' customers. DOHS and DPUC also ruled that Greenacres' customers would be billed at their old rate until all the residences were metered. BHC could then bill them at the same rate it charged other customers in the area. BHC was given a schedule for improving the Greenacres system. It also was required to submit certain financial information to the DPUC and to notify Greenacres' customers of the acquisition.

This ownership transfer was facilitated by Connecticut's takeover statute, which empowers the state to promote system acquisitions as a way of correcting the problems of nonviable systems.

---

government, which must develop a plan to correct system deficiencies and mitigate health problems. Another form of takeover exists through the right of eminent domain or condemnation power of local governments. This authority has been exercised in cases where a purveyor consistently provided unsafe or unreliable water service.

Contract Service

The proper operation of any utility requires qualified professionals. A contract service program enables a restructuring agent (such as another utility or a private contractor) to provide professional support to existing or new systems at

---

BOX 5-5 Contract Service for Full Operation: Beckham County, Oklahoma

The Beckham County Rural Water District No. 2 board of directors in western Oklahoma was having a difficult time keeping up with the technical demands of system operations. The system's single employee lacked sufficient skills to operate the system, and when he quit in 1993, the board decided to seek outside help.

The board contracted with Water Systems Management (WSM), the for-profit subsidiary of the Oklahoma Rural Water Association, to fully manage and operate the district's 212-connection system. WSM looks after the system's 100-plus miles of water mains, its three wells, and its chlorination, storage, and pumping infrastructure. WSM also takes responsibility for meter reading, billing, accounting, and operation and maintenance. The water district provides material and equipment free of charge for WSM's use in operating the system.

This arrangement has proved to be cost-effective for Beckham County; contracting with WSM is less expensive than hiring a system operator.

---

a cost-effective level without the small system having to find, hire, or supervise its own personnel (see Box 5-5). The service contract establishes the frequency, duration, cost, and specific responsibilities being hired out. Such responsibilities may include routine system operation and maintenance, periodic performance monitoring, required water quality monitoring, wholesale purchasing, equipment maintenance, scheduled repair activities, on-call emergency assistance, utility billing services, or other tasks. Some of the major contract considerations in providing this type of service are as follows:

- *System improvements*: If improvements are needed to enhance reliability, safety, or water quality, then the restructuring agent must determine whether they are to be completed or scheduled for completion before providing service. The restructuring agent may also agree to accomplish the improvements as a part of the contract.
- *Access*: The contracted services may be limited to facilities located on property owned by the system or located where guaranteed rights-of-way, easements, or unrestricted access exist for servicing, maintenance, and repair work.
- *Expansion*: If the system's service area is expected to expand, the restructuring agent may want the option to either approve the expansion or discontinue its contract services.
- *System contact*: The system should designate an accessible, responsible individual as the official contact.
- *Term of service*: The contract should specify the length of service being provided. Provisions for extensions may or may not be included.
- *Legal authority*: The contract should specify which system representa-

tives are authorized to contract services, commit to expenses, and be held accountable. Some restructuring agents require a hold harmless clause, especially regarding water quality conditions.

Support Assistance

Support assistance may be provided by a restructuring agent to a troubled water system on either a one-time or a continuous basis. The assistance may include operator training, information system support, purchasing of equipment and supplies, development of computerized mapping or infrastructure data bases, financial management or grant procurement assistance, or technical and engineering expertise. The major policy challenge is usually determining charges that will compensate the restructuring agent fairly without crippling the system seeking assistance.

Support assistance may take any of several forms. A *joint operating agreement*, for instance, can benefit two or more utilities that have complementary facilities, skills, or other assets; ideally, the strengths of each system will help correct the deficiencies of the others. This contractual relationship may include the sale or sharing of a portion, such as supply or storage, of a major facility. Detailed cost-sharing and responsibility assignments should be specified in the agreement.

Mutual aid agreements are likely to be between utilities of similar size and circumstances; any fees involved are usually low. An example might be two or more systems that join in making volume purchases to get volume discounts on supplies or water. Other examples are the sharing of equipment to handle special circumstances and the joint purchase of technical support programs for operator training.

Restructuring Agents

Any number of organizations or private or public utilities may serve as restructuring agents. Most fall into one of four categories: (1) nonmunicipal nonprofit organizations, (2) regional water authorities, (3) urban governments, and (4) investor-owned utilities.

Nonmunicipal/Nonprofit Organizations

This group is typified by the country's numerous rural water associations and rural electric cooperatives. Since the 1930s, these organizations have been empowered to provide utility services to rural entities. In most cases, these organizations are governed by a board of directors elected from the association's

---

BOX 5-6 Combined Management of Water, Wastewater, Electric, and Telephone Utilities in Arizona

The tribally owned Tohono O'Odham Utility Authority (TOUA) in southern Arizona manages all utility services for a reservation approximately the size of Connecticut but having a population of less than 10,000, as compared to Connecticut's population of 3 million (Rural Electric Research Project, 1994). The TOUA provides electric, water, wastewater, and telephone services to all residents of the Tohono O'Odham Nation. In total, the TOUA oversees 51 water systems serving 52 villages.

Central administration of all the utilities provides considerable economies in overhead and staffing costs and allows customers to do "one stop shopping" for their utility needs. The billing department provides one bill per customer covering all utilities. The TOUA staff read water and electric meters at the same time, share expensive equipment (such as backhoes and trenchers) and maintain it in a central facility, and have a mapping system that covers all utilities.

---

or cooperative's membership, which usually includes only the utility's customers or shareholders.

The National Rural Electrification Cooperative Association (NRECA) and the Electric Power Research Institute have recently advocated the movement of cooperative power utilities into the water works industry. This is a logical extension of services, particularly in rural and nonmetropolitan areas, where many of the staffing requirements, metering services, equipment demands, and service policies for power and water customers are similar (see Box 5-6).

In 1994, NRECA and the National Rural Utilities Cooperative Finance Corporation organized a joint task force to study the need and roles of rural electric systems in rural water and wastewater business. The task force's final report, *Community Involvement Opportunities in Water–Wastewater Services*, outlines different activities a rural electric system might be able to assume and an overview of the various issues involved in working with rural water and wastewater systems.

Regional Water Authorities

Regional water authorities may be composed of a consortium of several water purveyors or a single municipal or county government with territorial responsibilities. A classic example of this exists where a large municipal utility provides wholesale and retail service to customers throughout a metropolitan and rural area. Another example is the authority provided to public utility districts in Washington State; these districts have countywide authority and taxing capability to provide a range of utility services, including water, for county customers. These authorities generally focus on addressing comprehensive regional issues and aiding small water systems throughout the area. This often involves satellite management, because of the remoteness of many locations and the difficulty of providing direct interties.

Urban Governments

Larger individual municipal government utilities with urban levels of service also provide small system assistance. As growth and expansion widen the municipal boundaries and service territories of large cities or metropolitan governments, these entities are technically, managerially, and financially poised to take over responsibility for small systems. As government agencies, they can often obtain grant and loan funds for small privately owned water systems that would otherwise be ineligible. Frequently, however, customer concerns regarding other government policies, rate impacts, and city-county disputes over annexation of unincorporated areas affect the acceptance of these services.

Investor-Owned Utilities

Investor-owned utilities are economically motivated to provide assistance to small systems in cases that are profitable. As a group, these utilities develop and practice a "business plan" approach to utility service. In most cases, state public utility commissions regulate investor-owned utilities, so their level of service, pricing structure, and accounting and recordkeeping practices are closely scrutinized and regulated. Nonetheless, the economies of scale and the entrepreneurial expertise of investor-owned utilities make them effective candidates for small system assistance.

Barriers to Restructuring

Various factors can frustrate attempts to restructure small water systems:

• *Physical condition of system*: The most deteriorated systems are the most difficult to upgrade and finance. Who bears responsibility for evaluating and paying for system improvements is a key policy issue.
• *Location*: Remote locations make satellite service or direct interties more expensive and difficult and complicate routine operation and maintenance, as well as emergency support.
• *Density*: Operating and capital costs are more easily absorbed by systems with high densities rather than those with sparse populations spread over a large service territory.
• *Data*: Any serious lack of system, operational, or financial data makes it

difficult for a restructuring agency to accurately assess what will be involved in assisting a system.

• *Regulation*: Inflexible interpretation or implementation of regulations discourages restructuring agents from providing assistance. For instance, an expectation that a utility immediately comply with water quality standards with a subpar system might discourage them from purchasing it.

• *Politics*: Local political issues, including voter resistance to water system ownership changes and budget conflicts, may discourage some potential restructuring agents from active involvement in restructuring if such issues lead the restructuring agent to fear political resistance to its efforts.

• *Finances*: The lack of government funds to promote feasibility or regionalization studies often delays restructuring. In addition, the lack of funding assistance to correct or mitigate existing problems places the entire burden of implementing costly improvements on the restructuring agent that takes over the failing system. The inability of investor-owner utilities to access government financial assistance may further minimize their involvement.

• *Liability*: Liability risks, either real or perceived, assumed by restructuring agents that take over failing systems discourage restructuring activity. This is particularly true in cases where a system is in noncompliance with water quality standards and the restructuring agent cannot reach agreement with regulators about a reasonable compliance schedule.

• *Control*: The loss of decisionmaking powers, control, or ownership often leads systems in trouble (or the communities the systems serve) to reject efforts to restructure.

• *Growth impacts*: Given the perception that better-managed or upgraded water utilities can encourage growth, concerns about such growth may create opposition to restructuring.

• *Water resource allocation*: Currently, interpretation of water rights law, particularly in the western states, makes it cumbersome or impossible to change either the place of use or purpose of use in transfers of water rights. Appropriate state legislation can correct this situation by allowing such rights to be transferred to restructuring agents that have the ability to provide a supply of water through direct interties.

To reduce or eliminate these barriers and stimulate the use of more restructuring activities, several changes need to be made. The following suggested list covers a range of procedural and financial incentives for restructuring; these incentives can be created at the federal, state, or local levels.

• As mentioned previously, the federal government should provide SRF monies only to states with public health performance appraisal programs for water systems. The SRF monies should be used in part to provide funds to organizations and utilities that can participate in restructuring, both for regional and system-specific feasibility studies and for water system improvements. SRF funds should also be available to states to provide technical and administrative support to local jurisdictions involved with restructuring and satellite management.

• Federal and state agencies should provide low-interest loans and grants to public and private entities that can assist with restructuring unsustainable water systems.

• Such loans and grants should not be used to prop up unsustainable systems.

• Both federal and state governments should provide tax incentives to investor-owned utilities to assist with restructuring unsustainable water systems.

• Congress should remove the "no credit elsewhere" test (a requirement that a system have exhausted conventional lending options) from the Rural Utility Service (RUS) grant and loan program; at the same time, water systems should be eligible for RUS grants and loans only if they have completed a public health performance appraisal. In cases where more expensive financing is available from a private lender, RUS grants and loans are currently provided only when the resulting rates to water users would not be comparable to those in nearby areas.

• Congress should modify section 312 of the Rural Electrification Act to enable electric utility borrowers to invest more than 15 percent of their total utility plan in nonelectric activities. This would allow rural electric utilities to participate more freely in water system projects.

• The federal tax code should be changed to ensure that RUS water grants or loans are not considered nonmember income for purposes of federal income taxes, lest their contribution bring total nonmember revenues above 15 percent of total revenues, which triggers tax consequences on the utility.

• State public utility commissions should allow adjustments to the rate base of utilities to reflect the cost of acquiring a failing system.

• State public utility commissions should allow restructuring agents to depreciate systems for which they have assumed responsibility.

• Federal and state governments should provide temporary waivers to restructuring agents for liabilities associated with SDWA violations in cases where the restructuring agent has acquired a failing water company. These waivers should be tied to reasonable compliance schedules and activities.

IMPLEMENTING CHANGES TO PROMOTE SUSTAINABILITY

The ability to successfully implement changes is critical to maintaining sustainability for small systems with limited resources. Leadership is vital to this effort. Communications skills in particular can prove critical to focusing public debate constructively. As Bennett (1993) notes in *Managing the Human Side of Change*, "During any change process, something new begins only after something else ends. It is the loss associated with this ending that people seek to avoid. In other words, people don't fear change—they fear loss." Good communication skills can help a debate focus less on what is being lost than what is being gained.

The appropriate strategy and tactics to minimize this fear of loss and develop a viable solution depend on the missions and characteristics of the restructuring agent and the small system being assisted. Bryson and Delbecq (1979) listed six steps to identifying and effectively implementing strategies. These include (1) initial agreement concerning the purpose of the action, (2) needs assessment, (3) search for possible solutions, (4) proposal development, (5) proposal review and adoption, and (6) implementation.

Bryson and Delbecq found that in politically difficult situations, the importance of giving good attention to the initial agreement (step 1) and of the proposal review and adoption phases (step 5)—two of the more political phases—increased. They found the initial agreement to be the most important phase in politically difficult situations and the least important in politically easy situations.

Increased technical difficulty, in the absence of increased political difficulty, tends to increase the relative importance of the needs assessment and search for possible solutions phases—two of the more technical phases. Interestingly, the identification of more difficult solutions (step 3) tended to increase the relative importance of the initial agreement and the proposal review and adoption phases. In other words, increased solution difficulty has something of the same effect as increased political difficulty and requires more attention to the political steps. Demonstration projects may be critical, either in solving technical questions or in generating political support for the proposed solution.

Finally, Bryson and Delbecq found that in most cases, a final solution is not so much "planned" as it is negotiated and haggled out—preferably in a structured, goal-oriented fashion through an exercise of leadership and appropriate problem-solving processes. In such debates, local involvement is essential to ensure that if supporting agencies providing assistance end their involvement, the local community will have the capacity to sustain the system or seek an alternative solution (Okun and Lauria, 1991).

CONCLUSIONS

Water supply systems of all sizes face increasing challenges as they attempt to meet customer expectations, health requirements, and safety considerations within affordable rate structures. The problem is particularly pressing for smaller systems, which generally do not have the resources needed to implement necessary changes.

- **The solutions to the problems of small water supply systems should focus on ensuring sustainable water supply service.** A sustainable water supply system is one that has a commitment to meet service expectations; has the technical, managerial, and financial capacity to meet public health and safety performance requirements on a long-term basis; and has access to an adequate source of water.

- **Meeting the escalating public health regulations and customer demands for water service requires that small systems use a business planning approach to assess their options.** The local community should be involved as much as possible to ensure community endorsement.

- **In some cases, the problems of small systems will best be solved through restructuring, that is, by relinquishing control of some or all aspects of the water system to another organization, known as a restructuring agent.** Options for restructuring include direct transfer of system ownership to the restructuring agent, receivership or regulatory takeover, contracting services for some system components while leaving system ownership unchanged, and provision of technical support on a regular basis.

- **A variety of barriers can discourage efforts to restructure small water systems.** Barriers include disputes over who should pay for system improvements, lack of data for assessing what will be involved in assisting a system, requirements that restructuring agents be held liable for violations of drinking water standards by the small system, political resistance to ownership changes, lack of funds to promote feasibility studies, and water resource allocation policies. Greater efforts are needed to overcome these barriers.

RECOMMENDATIONS

- **States should establish programs requiring all water systems to conduct public health performance appraisals.** Only systems that have passed a performance appraisal should be issued an operating permit. Those that do not receive permits should be obliged to restructure.

- **The federal government should provide state revolving fund monies and rural utility service grants and loans for drinking water systems only to states with official public health performance appraisal programs.** This will ensure that federal funds are not used to prop up unsustainable water systems. States should be able to use SRF monies to help develop their performance appraisal programs.

- **Federal, state, and local governments should develop incentives to encourage the restructuring of unsustainable water systems.** Incentives include providing SRF monies and tax breaks to restructuring agents that take over failing small water systems, allowing adjustments to the rate base to reflect the costs of acquiring a failing system, allowing restructuring agents to depreciate systems for which they have assumed responsibility, and providing temporary waivers to SDWA requirements in cases where a restructuring agent has acquired a failing water company. Waivers to SDWA requirements should be tied to reasonable compliance schedules.

REFERENCES

AWWA (American Water Works Association). 1991. Water Rates, Manual No. 1, Fourth Edition. Denver: AWWA.

AWWA. 1992. Alternative Rates 1992, Manual No. 34. Denver: AWWA.

Bennett, M. W. 1993. Managing the human side of change. In Proceedings of the American Water Works Association/Water Environment Federation Joint Management Conference. Denver: American Water Works Association.

Bryson, J. M., and A. L. Delbecq. 1979. A contingent approach to strategy and tactics in project planning. Journal of the American Psychological Association (April):167–179.

Campbell, S., B. Lykins, Jr., and J. A. Goodrich. 1993. Financing Assistance Available for Small Public Water Systems. Denver: American Water Works Association.

Cromwell, J. E., III. 1994a. Strategic planning for SDWA compliance in small systems. Journal of the American Water Works Association (May):42–51.

Cromwell, J. E., III. 1994b. Generic elements of a state viability program. Presented at the AWWA National Conference, New York, June 19, 1994.

EPA (Environmental Protection Agency). 1990. Improving The Viability of Existing Small Drinking Water Systems. EPA 570/9-90-004. Washington, D.C.: EPA.

McCall, R. G. 1986. Institutional Alternatives for Small Water Systems. Denver: American Water Works Association.

Okun, D. A., and D. T. Lauria. 1991. Capacity Building for Water Resources Management. New York: United Nations Development Programme.

Okun, D.A. 1995. Addressing the problems of small water systems. In International Water Supply Association Congress, Burban, Republic of South Africa. London: International Water Supply Association.

Rural Electric Research Project. 1994. Rural Water/Wastewater Study, Volume 2: Case Studies and Management Issues. Washington, D.C.: National Rural Electric Cooperative Association.

Wade Miller Associates. 1991. State Initiatives to Address Non-viable Small Water Systems in Pennsylvania. Arlington, Va.: Wade Miller Associates.

WADOH (Washington Department of Health). 1994. Small Water Utilities Financial Viability Manual. Seattle: WADOH.

6

Training Operators for Small Systems

Competent operating personnel are vitally important to the sustained, safe operation of small water systems. Accordingly, good operator training is as essential to improving small water systems as are improved technologies, organizational fixes, or regulatory oversight. Without adequately trained personnel, even a well-financed and organized system with the most advanced technology and regular compliance visits will fail to reliably deliver safe drinking water to its customers.

Unfortunately, the training available to most small system operators falls far short of meeting their needs. This training deficit became especially evident in the early 1990s, when violations of the Safe Drinking Water Act (SDWA) began accumulating in the U. S. Environmental Protection Agency's (EPA's) Federal Reporting Data System. Between 1989 and 1994, the number of community water systems violating the SDWA increased by a third, from 12,295 to 16,779, and the number of violations they incurred more than doubled, from just over 40,000 to over 88,000.

The mounting number of violations made it clear that many small system operators found it difficult to comply with the increasingly complex regulations introduced by the 1986 SDWA amendments. Congress and others involved in passage of the SDWA and its 1986 amendments had assumed that operators either had or would easily acquire (presumably from existing state programs) the skills necessary to comply with the expanding regulatory requirements they faced. However, most small system operators come to their positions through circuitous routes, with relatively little formal training (see Box 6-1 for a typical example).

**BOX 6-1 Genesis of a Small System Operator:
A Typical Story**

This operator's first involvement with small water system operation began
shortly after her family moved to a small community located outside a medium-
sized city. Concerned about the community water supply, she volunteered to keep
the system's account records. Within a year, she became a volunteer water system operator. A few years later she found herself in the position of operator-in-
charge, bearing primary responsibility for system maintenance and repairs, water
source development, and treatment operations. Thus, in a very short time she had
vaulted from volunteer bookkeeper to operator-in-charge, with little qualification
other than her interest in the quality of her community's drinking water. Character-
istically, she received little formal training to prepare her to undertake the significant public responsibilities of her new position.

The states have struggled to meet the training needs of these small system operators.

LIMITATIONS OF EXISTING TRAINING PROGRAMS

The reasons for the training deficit among small system operators are many.
To begin with, the training available to small system operators is often provided
haphazardly through a mix of state-run workshops or seminars, informal instruction from state regulators during on-site inspections, and (in some cases) training
provided by technical schools, university continuing education courses, American Water Works Association (AWWA) courses, equipment vendors, or rural
water associations. These programs are not coordinated in any meaningful way.
In addition, the remote location of some small systems, the part-time or volunteer
status of most small system operators, and the cost of reaching and attending
training courses discourage these operators from taking advantage of these resources.

Perhaps more important, most water treatment operator training programs
are designed for operators of medium and large systems and thus fail to give
small system operators the combination of broad general knowledge and hands-
on practical training they need. Most courses provide general training of a depth
that goes beyond what a small system operator will ever require, yet skip many
operational basics.

For instance, many beginner training classes include extensive sections on
complex water treatment processes that few operators of very small systems will
ever use. Similarly, most operator training programs (as well as many state
certification requirements) cover general technical aspects of numerous water
treatment technologies, some of them quite advanced, while operators of smaller

systems need specific, hands-on training in only the one or two relatively simple treatment technologies their systems use. Thus, the operator of a system that uses a ground water well and a simple disinfection technology such as chlorination may be trained and tested in multiple advanced technologies appropriate only for larger, surface water systems. Meanwhile the course will offer no training in the challenges the small system operator often faces, such as how to fix a chlorinator, how to take monthly samples without contaminating them, or how to fix and refill a water main without contaminating the entire system.

Further aggravating the situation is a lack of consistent training standards and certification requirements across the nation, which makes it difficult for small system managers or operators to even determine if the operators need training. Educational requirements for small system operators can be quite minimal; some states do not even require that operators have high school diplomas. Most operators appear to be particularly undertrained in management and administration. Good management and administration are as essential to sustained system operation as are treatment and distribution issues, but most states have no established minimum requirements regarding these management and administration issues, and most training and testing programs ignore them.

All of these problems seriously compromise training of the thousands of small system operators across the country, especially for operators of the very small community systems (those serving fewer than 500 people). For them in particular, the available training usually fails to be either comprehensive or specific enough to meet their needs. This training shortfall constitutes a major threat to the safe and sustained operation of the nation's small water systems and to the ability of those systems to meet the standards of the SDWA. The rest of this chapter will look at the network of present training resources and examine how they can be improved and coordinated to provide more adequate training for small system operators—a task in which the EPA can play a key role.

State Programs

Operator training and certification activities have been an integral part of state drinking water programs for many years. All 50 states currently have operator certification regulations in place. According to a 1995 survey conducted by the Association of State Drinking Water Administrators (ASDWA), 49 of the state programs are mandatory. Forty-nine of the states require operators to pass an exam, 50 require experience, and 47 have educational requirements in place. Forty-six of the states require that certificates be renewed every 1, 2, or 3 years. In addition, 37 states currently require operators to obtain continuing educational credit for certificate renewal. Some states exempt operators of very small systems from some or all of these requirements. In general, state certification requirements for small system operators tend to suffer from the problems described earlier—a failure to be specific to the needs of small system operators, lack of

---

BOX 6-2 The Illinois Water Operator Certification Program

The Illinois water operator certification program is in many ways a typical state program. State regulations require all public water supply systems within the state, except for a few state-owned or -operated facilities and certain communities that buy treated water from another supply, to have a certified operator. Four levels of certification may be achieved by acquiring varying amounts of appropriate experience and passing standardized tests. The certification levels increase in difficulty from distribution system operation only (class D) to the lime softening and filtration of surface waters (class A). The minimum educational requirement is the completion of grammar school, with experience credit granted to those with higher levels of education.

Like tests in most states, Illinois certification tests determine competency primarily in the technical aspects of treating, testing, and distributing potable water, with some attention paid to regulatory knowledge, recordkeeping, and reporting. Little administration or management knowledge is tested.

---

emphasis on areas of expertise other than treatment and distribution, and inconsistency in educational requirements. Box 6-2 describes a fairly typical operator certification program, that of the state of Illinois.

In an attempt to supply operators with the knowledge necessary to become certified, the states have taken the lead in training small system operators. In fact, state training programs supply virtually all of the training that operators of the smallest systems receive, yet states are short of the personnel needed to conduct adequate training. According to preliminary analyses of a resource needs model developed by the ASDWA and the EPA, states employed approximately 2,160 full-time employees, or full-time equivalents (FTEs), in fiscal year 1992 for drinking water program activities. Of that number, 76 were dedicated to training and development activities and 82 were dedicated to operator certification activities, for a total of 158 FTEs. The model projected that by fiscal year 1995, states would need 113 FTEs for their operator certification programs, 42 FTEs for operator certification assurance activities, 78 FTEs for ongoing training activities, and 34 FTEs for rule-specific training, for a total of 267 FTEs.

States have developed a number of tools to train and certify operators. These include newsletters informing operators of changes in state and federal regulations; information mailings and documents related to specific rules; training seminars and workshops provided directly by the state; state participation in and notification of workshops and seminars provided by other organizations, such as the AWWA and National Rural Water Association state affiliates; and certification exams held around the state. In addition, many states review the status and capabilities of operators as part of their state sanitary survey processes. During

---

BOX 6-3 Other Training Resources

In addition to the states, the National Training Coalition, and the National Environmental Training Center for Small Communities, several other sources also presently deliver some training to small system operators.

The National Rural Water Association (NRWA) is the national body of state rural water associations, with a primary mission to train and assist small water system professionals on matters of drinking water and wastewater treatment. Most of the training is done by technical "circuit rider" staff who make on-site technical assistance visits. However, staffing on these associations is usually limited to no more than one circuit rider per state, so they cannot provide training in any depth for operators of the smallest systems.

The American Water Works Association (AWWA) offers training programs, but these are aimed at operators of more advanced and larger systems and are usually too expensive, inaccessible, or technically specialized or advanced for small system operators.

Technical schools, community colleges, and university extension services offer additional sources of training; these are typically used by operators of larger small systems (those serving between 3,300 and 10,000 people). Such training usually takes the form of a continuing education class rather than a comprehensive training course. These courses generally meet only a narrow need, and (partially because they usually require a tuition payment) tend to reach only the most aware and motivated operators or those who work for larger systems inclined to pay tuition.

Equipment vendors sometimes provide training, either onsite or in seminars organized through NRWA state affiliates, AWWA sections, or state initiatives. But as might be expected, these courses are usually limited to the equipment sold by the vendor in question and rarely cover areas beyond treatment and distribution.

---

these inspections, state staff visit the water utilities and work with operators to answer questions and provide additional training.

As noted earlier, these programs constitute the sole training for most small system operators. In tandem with the national programs described below, these state programs make a logical starting point for creating a more coordinated and comprehensive small system operator training network. (These state efforts are also supplemented by training programs from schools, associations, and vendors, which could also play important but less central roles in a more coordinated network; those programs are described in Box 6-3.)

National Programs

On a national level, two recently created groups now seek to address operator training inadequacies; both could play roles in creating a more coordinated, comprehensive training network.

The National Training Coalition

The primary goal of the National Training Coalition (NTC) is to promote cooperative work, through its six member groups, to help develop state training coalitions (NTC, 1995a). The NTC's members include the AWWA, ASDWA, the EPA, the National Environmental Training Association, the National Rural Community Assistance Program, and the National Rural Water Association.

In 1991, the NTC conducted a national survey of state and federal drinking water regulatory agencies and held two regional fact-finding workshops. The survey and workshops revealed that while most states had active regulatory training programs, there was little formal coordination and cooperation among the organizations, agencies, and institutions providing the training. The NTC found that this lack of coordination and cooperation created a duplication of efforts, conflicting training for operators, and confusion regarding proper procedures and regulatory interpretation. The survey found that both state and federal drinking water administrators felt that existing training efforts were not fully meeting the needs of water system operators (NTC, 1995a).

In a follow-up survey conducted by the NTC in 1994, the majority of respondents indicated that the main focus of training programs throughout the country was to provide information on future drinking water regulation. This finding suggests that a priority has been placed on regulatory compliance at the expense of operational performance in most training programs. The survey further revealed that the main needs of training providers include funding, training resource materials, and adequate staff devoted to the effort (NTC, 1995b).

The National Environmental Training Center for Small Communities

The National Environmental Training Center for Small Communities (NETCSC) was initially established to support the EPA's administration of the Clean Water Act. The center's mission was subsequently expanded to include drinking water and solid waste management. The stated mission of the NETCSC is to support environmental trainers in their efforts to improve the quality of wastewater, drinking water, and solid waste services in small communities.

In 1994, the NETCSC conducted a survey of small community environmental training experts concerning issues and trends in small community environmental training. The survey's report, *Small Community Environmental Training: Trends-Issues*, reveals 25 trainers' perspectives on the status of environmental training in small communities. According to nearly half the survey respondents, the people who most need training often do not receive it. One respondent noted that the logistics of providing training to 57,000 small community systems across the nation are overwhelming (NETCSC, 1994). A small community's ability to afford needed training was highlighted as a concern by two-thirds of the survey respondents.

PROPOSED IMPROVEMENTS FOR TRAINING PROGRAMS

To provide better training for small system operators, it is first necessary to examine the tasks such operators normally perform. Typically these operators must perform a wide assortment of work (see Box 6-4). It has long been widely believed that a small system operator must exhibit competence in only two broad technical areas, treatment and distribution, for a system to be viable. However, while these two areas are important, a small system operator also needs to be competent in administrative, financial, customer service, and other skill areas. A good operator is a "jack of all trades."

Table 6-1 presents a task analysis of eight typical small system operator skill areas. Although it is beyond the scope of this chapter to establish specific benchmarks for training competency, the table illustrates the considerable depth and breadth of knowledge a small system operator might require. The type and depth of knowledge needed, of course, vary considerably according to the size and nature of the small system, which by definition might serve a population as small as 25 people (15 connections) or as large as 10,000. The water consumption associated with these systems would vary from less than 2,500 gal per day to close to 2,000,000 gal per day, while the number of personnel required would vary from one part- or full-time operator to a staff of 12 or more.

In 1991, the Association of Boards of Certification recognized the importance of the additional areas of training through the establishment of "Need to Know" criteria. The association has begun to use the criteria in standardized tests for small and very small system operators in three states.

BOX 6-4 A Small System Operator's Typical Day: December 12, 1995

Because Joe is a township supervisor and township employee, as well as the local water system's certified water treatment plant operator, this winter day began very early. His first duty, at 4:00 a.m., was plowing snow along township roads for school bus transit and other traffic; at one point his truck lost its brakes, causing him to have to ditch the truck to stop it. It was 9:00 a.m. before he got the truck out and could get to his other duties.

At that point he was ready to check his water treatment plant. He began by checking the chlorine monitor to see if it indicated a satisfactory chlorine residual at the entry point to the water distribution system; it did. He then noted the turbidity of the surface water source, checked the flow of the raw water source at the water meter and recorded these read-outs in the daily record book.

After that, Joe checked the treatment plant's chlorinator and solution tank, and noted in the log that the pump was working properly and the solution tank nearly full. He then performed calibration checks on chlorine and turbidity. Finally, he ran his routine checks on the distribution system, which took less than a half-hour.

TABLE 6-1 Small System Operator Task Analysis

General Work Area	General Work Item	Specific Work Items	Associated Knowledge Desired
Source and supply	Selection	Economics Quality Adequacy	Operational costs and manpower requirements, economic evaluation, recordkeeping and reporting requirements, water quality analysis and interpretation, contamination risk
	Maintenance	Protection of quality and quantity	Withdrawal and source protection regulations, reservoir management, zoning, quantity assessment, quality analysis, treatment
Treatment	Process selection (including point-of-entry devices)	Economics Reliability Ease of operation Performance	Economic evaluation of operation costs, history of process performance and operation requirements, contingency plan if initial process fails
	Operation and maintenance	Continuous Intermittent Chemicals and chemical handling Laboratory and quality control Automation Emergency response Preventative maintenance Inventory	Staffing requirements and planning, process performance under differing conditions, chemistry of chemicals employed, safety, laboratory analysis, quality regulations, automated control and monitoring of system operation, computer applications, recordkeeping and reporting regulations, emergency response planning and implementation, contingency planning and implementation, customer protection procedures,

Distribution		
		maintenance skills for equipment installed, maintenance planning, inventory planning and control
Operation and maintenance	Flushing	Flushing plan, fire-flow rate determination, water quality determination, traffic control, equipment operation, trench safety, repair procedures, repair disinfection, customer protection (boil order issuance) procedures, system data base, emergency response planning and implementation,
	Repair of service mains, valves, and hydrants	
	Installations of new service mains and services	
	Emergency response	contingency planning and implementation, hydraulic modeling, inventory planning and control, system performance and condition regulations, leak detection and location methods, recordkeeping and reporting, storage tank operation and maintenance, lead and volatile organic chemical paint regulations, booster station operation optimization
	Automation	
	Inventory	
	Storage	
	Booster stations	
	Cross-connection control	
Maps and records	"As built" system maps	Drawing interpretation and preparation, filing and retrieval, main and service line locations and marking
	System performance	
	Location requests	

continued on next page

TABLE 6-1 *Continued*

General Work Area	General Work Item	Specific Work Items	Associated Knowledge Desired
Meters	Justification	Expense allocation	Cost-benefit analysis, billing procedures, water loss accountability, water conservation methods
	Reading	Economics Manpower availability	Reading procedures, reading equipment
	Operation and maintenance	Economics Installation Maintenance Replacement	Meter installation, meter testing, meter repair, meter tracking
Customer service	Customer needs	Complaint/inquiry investigation and solution Emergency response Customer accounting and collections Customer expectations determination New customers	Customer interaction skills, remedy alternatives, customer accounting methods and procedures, rules of service, collection procedures, survey techniques, new customer customer establishment
Finance	Payroll	Hourly and salaried employees	Time accounting, work classification and rate, expense allocation

Accounts payable	Materials Supplies Utilities Services Damages	Purchase procedures, payment procedures
Accounts receivable	Sold inventory or surplus Sold services Damage to facilities	Sales procedures, collection procedures
General ledger	Chart of accounts Assets Liabilities Income accounts Expense accounts Equity accounts	Intermediate financial accounting, recordkeeping procedures, state regulation (if applicable)
Funds acquisition	Acquiring loans Acquiring grants	Basic financial accounting, tax regulations, state regulations, loan and grant programs
Private company requirements	Depreciation Common income Accruals Taxes	

continued on next page

198

TABLE 6-1 *Continued*

General Work Area	General Work Item	Specific Work Items	Associated Knowledge Desired
Human resources	Staffing levels	Determine proper levels	Task analysis
	Staffing qualifications	Determine necessary job qualifications	Determine optimum match of job and employee qualifications
Administration	Risk control	Safety and environmental audits Risk analysis Insurance Risk control	Insurance level determination and purchasing, risk analysis, risk reduction
	System performance analysis	Performance benchmarks Performance goals and monitoring	Determination of industry performance benchmarks, setting and achieving performance goals
	Planning	Capital Growth Expense Revenue Emergency	Strategic and long-range planning of all business factors

Two general relationships characterize how system size affects necessary operator expertise. First, as the size of the system increases, and with it the number of employees, the number of job areas in which each employee must be proficient decreases. Second, as the size of the system increases, the depth of knowledge in each individual's focus area usually increases. In other words, small system operators require basic knowledge of all the job areas (from source and supply to administration), while the more specialized (and numerous) operator-employees of large systems must each possess detailed knowledge of a narrower range of skill areas.

The ideal training program for any size system, therefore, would divide the skills into a number of areas, each with several skill layers, each layer more advanced than the previous. This approach could be visualized as a training matrix such as that shown in Table 6-2. On the vertical axis are skill areas, and on the horizontal axis are the levels (depth) of accomplishment or difficulty. The operator of a small system would generally train in many areas (vertical alignment), while employees of a larger system would tend to train more vigorously in fewer areas (horizontal alignment).

The training matrix in Table 6-2 could be used as a starting tool to organize and detail the skills that must be mastered for safe and sustainable plant operation of systems of different sizes. It is important that no matter what the size of the system, the entire range of skill areas be covered. As mentioned before, training typically has addressed primarily distribution and treatment concerns, without considering the wider, more general requirements for system sustainability. A more comprehensive approach will be more successful in the long run. Training geared towards improving skill areas that are presently underemphasized (meters, customer service, financial, administration, and human resources) will greatly benefit most small systems. Future training programs should use a comprehensive approach covering all areas at depths appropriate to each system's size.

State agencies could play an important role by requiring operator, management, and system certification. The system certification could be linked to the annual operating permit discussed in Chapter 5. In establishing certification standards, states should recognize that the full complement of identified skills are necessary to provide a sustainable operation. Failure to provide competency in all areas may, in many cases, lead to an unsustainable system and eventual failure to protect public health through compliance with the drinking water regulations. Any certification requirement should therefore include not only the technical operation skills, but also the other important skills identified. Competency in these areas could be proven through state-conducted performance assessment testing, successful completion of required courses, other advanced certifications that include the necessary content (for example, a professional engineering or accounting degree), or assignment of the work to a qualified professional.

The technical treatment competency requirements for a small system operator should be limited to a basic understanding of public health concepts, appli-

TABLE 6-2   Training Matrix: Examples of Training Levels

Skill	Level 1	Level 2	Level 3
Source of supply	Quality tests, regulations	Watershed survey	Economics and long-range planning
Treatment	Specific process used, regulations	Other treatment processes	Economics and process selection
Distribution	Main repair, regulations	Pump repair, control systems	Hydraulic modeling
Meters	Meter reading	Meter repair	Automatic meter reading, economics
Customer service	Answering customer questions	Scheduling customer service activities	Determining customer desires
Financial management	Billing, collections	Payroll, general ledger	Rate design, borrowing
Administration	Sustainability	System growth	System operation as a business, lobbying for favorable legislation
Human resources	Safety	Hiring, firing, employment regulations	Total quality management concepts

cable regulations, source protection, distribution system operation, and competency only for the actual treatment process(es) the operator's system employs. It
does little good to require small system operators to learn processes they will
likely never use; indeed, such a requirement actually may discourage such operators from seeking certification. At the same time, however, small system operators will need to have some knowledge of how the particular treatment process or
processes used in the plant function. An understanding of the principles by which
the process operates is important for responding to system malfunctions; knowing how to turn handles and record meter readings is not enough to recognize and
address operating problems. If the operator does not have sufficient understanding of the treatment process, then such expertise must be available to the operator
on short notice from a circuit rider, consultant, larger water utility, state agency
representative, or another source.

   Illinois already uses this type of system-specific certification program for
industrial wastewater treatment plant operators. The state issues a "class K"
certificate valid only for specific plants; the examination tests the applicant's

knowledge of principles, techniques, permit requirements, math, and safety, in addition to requiring a flow schematic of the applicant's particular plant. The test is geared towards testing the applicant for knowledge crucial for operation of his or her particular plant and certifies the operator for only that plant (Illinois Environmental Protection Agency, 1983).

The Illinois system-specific technology certification program may provide a good model for treatment certification programs in other states. Of course, advanced treatment certifications covering many treatment processes should remain available for larger system operators, more ambitious small system operators, and circuit riders exposed to a wide and varying assortment of processes and situations.

IMPLEMENTATION

Improving small system operator capabilities will require significant changes in the training available to those operators. Such changes will not occur without a well-orchestrated effort at the national level, including a substantial and continuing commitment of resources. Developing and delivering a comprehensive, nationally available training program will also require a leadership agency to guide the effort. This is a vital role the EPA can and should fulfill, and the Committee on Small Water Supply Systems strongly recommends that the agency do so. It is clearly within the EPA's mandate to provide technical assistance in support of the goals established by Congress under the SDWA. The SDWA Amendments of 1986 authorized the expenditure of up to $15 million per year to provide technical assistance to small systems struggling to comply with the requirements of the act. The EPA currently provides $6.5 million annually to the National Rural Water Association and the Rural Community Assistance Program for technical assistance to small water systems. However, this funding has not resulted in the development of coordinated training programs for improving the knowledge and skills of small water system operators. Historically, the EPA has had to concentrate on regulatory compliance at the expense of developing programs for improving operator performance. Focusing on operator performance may more efficiently address the fundamental issue, which is improving water supply.

To fulfill this recommendation, the EPA should

- establish an organizational work unit, based at headquarters, responsible for identifying desirable knowledge and skills for the successful operation of all aspects of drinking water systems;
- arrange for the development of multimedia tools to effectively deliver the needed training to system owners and operators across the country; and
- support efforts to coordinate and deliver training programs to operators of

all system sizes and types and in dispersed geographic locations. (This training would not include regulatory compliance, since that will vary from state to state.)

Obviously, these recommendations require a commitment of funds. Recognizing that new funding resources may not be available, this recommendation could be readily accomplished simply by reprioritizing resources already under the agency's control so that training receives as much emphasis as do development and enforcement of regulations.

In this role, the EPA should serve as a facilitator. Ideally, the new work unit would be staffed with experts in innovative training applications who, through the use of partnerships and contracts with experts in the field, would direct the development of interactive training modules in various formats for delivery to small system operators. In doing so, the EPA should draw first upon training materials that are already available through trade associations, public utilities, state programs, and others and are deemed to be of acceptable quality by a panel of expert advisors. Both the NTC and the NETCSC could be vital partners in this effort. These materials should be compiled and catalogued to determine where gaps exist and to establish priorities for creating additional training modules to fill those gaps.

Existing materials with acceptable technical content could then be submitted to a contractor with expertise in developing innovative training programs. The EPA would work with technical experts to develop acceptable training modules to fill areas where gaps exist before submitting those modules to the training program contractor. In creating these materials, the unit should draw on experience gained by some state programs and other training providers in the development of effective training programs; here again the NTC could lend valuable assistance. Training providers who are in direct contact with small system operators are well positioned to develop effective, interactive training programs that meet the needs of those operators.

Once the modules are completed for the entire program, the EPA would move into delivery mode. The agency should provide and promote access to the training modules as a national initiative. Electronic vehicles (video, computers, etc.) for dissemination should be fully exploited. To ensure that the training materials remain state of the art, the EPA should also have an ongoing program of technical review and modification to incorporate feedback from operators and technical experts regarding quality and relevance.

Such an effort could begin to fill the present gap between regulatory requirements and available training. A significant backlog already exists, for instance, with regard to developing the operator skills necessary to comply with disinfection, filtration, and corrosion control processes mandated by existing federal and state regulations. When developing new regulations, the EPA must, as a priority, identify operator skill requirements and training needs early in the process so that the appropriate training materials can be made available as the new regulations take effect.

The states should be designated as the lead agencies for delivering the national training program. The state agencies' organizations, locations, and professionally trained staffs enable them to effectively deliver the training programs. As designated lead training providers, the states would also be responsible for networking and coordinating with other training providers, such as state operator associations, state rural water associations, AWWA state sections, educational institutions, and others able and willing to support the effort. The EPA should vigorously promote such state leadership. Training must be given prominence in annual program grant funding and when negotiating the work plans for primacy program implementation. The EPA also needs to provide easy state access to the training materials and the training program development contractor to meet state-specific training needs.

The NETCSC could be the vehicle to institutionalize operator training into the EPA's mission and facilitate its implementation. The NETCSC operates under a cooperative agreement with the EPA. It has three primary functions: serving as the library for training programs and materials developed by state, association, and other providers; developing new multimedia training vehicles; and conducting train-the-trainer courses across the country. The NETCSC has developed several training programs covering key areas of water supply service; these training programs typically employ manuals, books, and videotapes. The center's Train-the-Trainer program could facilitate dissemination of these materials to a broader audience of operators across the country.

CONCLUSIONS

Good operation and management are fundamental to the sustainability of small water systems, yet operators of small systems often lack adequate training. Many regulatory violations and waterborne health risks could be avoided by an increased investment in operator training. Although efforts to train small system operators are beginning to improve, they have a long way to go:

- **Training is often unavailable or inaccessible for operators of very small systems.** The remote location of some small systems, the cost of traveling to and attending training courses, and the lack of personnel to serve as back-up operators discourage small system operators from enrolling in training programs.
- **The content of training programs often fails to meet the specific needs of small system operators.** Too often, training programs focus on theory and tasks for water treatment that only operators of larger systems need to learn in detail.
- **Training programs often fail to include the multiple key skill areas (treatment, distribution, source of supply, meters, customer service, finan-**

cial administration, and human resources) needed to ensure a viable system. Most existing training programs address only treatment and distribution.

• **Resources for supporting training programs need to be increased if operator training is to reach a satisfactory level.** Funding, program design, resource materials, and staff all need bolstering. The presently uncoordinated programs could benefit from initiatives like the NTC's efforts to facilitate the formation of state coalitions.

RECOMMENDATIONS

• **The EPA should guide the effort to improve training for small water system operators.** The EPA should build on the efforts of the NTC and the NETCSC and reallocate resources that implicitly overemphasize enforcement over technical training. Whether with new funds or through reallocated funds, the EPA should (a) establish an organizational work unit to identify appropriate operational knowledge requirements; (b) arrange for the development of multi-media training tools for nationwide delivery; and (c) vigorously support efforts to coordinate and deliver training programs in the field to dispersed operators.

• **State drinking water agencies should be responsible for delivering training programs developed by the EPA.** Training will be most effective if it is delivered locally to minimize the need for travel by small system operators.

• **Lead training agencies should prioritize each of the general key training areas and offer training accordingly.** States should establish competency benchmarks for all small system work areas.

• **States should rewrite their certification laws for small system operators to emphasize the processes employed by the certified operator's particular system.** Requiring more knowledge than necessary from a small system operator wastes his or her time and discourages the pursuit of certification. Operator certification should include classifications not only by system size but also for each of the general skill areas.

REFERENCES

Illinois Environmental Protection Agency, Division of Water Pollution Control. 1983. Procedures for Certification of Operators of Wastewater Treatment Works. IEPA/WPC/83-012. Springfield, Ill.: Illinois Environmental Protection Agency.

NETCSC (National Environmental Training Center for Small Communities). 1995. Small Community Environmental Training: Trends and Issues. Morgantown, W. Va.: National Environmental Training Center for Small Communities.

NTC (National Training Coalition). 1995a. Cooperation, Coordination, Communication, and Compromise... The Keys To Successful Coalition-Building. National Training Coalition Activity Report No. 2. Washington, D.C.: Association of State Drinking Water Administrators, National Training Coalition.

NTC. 1995b. Findings of the NTC. Washington, D.C.: Association of State Drinking Water Administrators, National Training Coalition.

Appendix

Committee on Small Water Supply Systems
Biographical Information

Vernon L. Snoeyink, who chaired the Committee on Small Water Supply Systems, was appointed to the civil engineering faculty at the University of Illinois in 1969. He is coordinator of the Environment and Science Program within civil engineering and was named Ivan Racheff Professor of Environmental Engineering in 1989. His primary area of research is drinking water quality improvement, and he has focused his efforts primarily on the removal of organic contaminants by activated carbon adsorption. In 1980, he coauthored the textbook *Water Chemistry*, published by John Wiley & Sons. He has been a trustee of the American Water Works Association Research Foundation and president of the Association of Environmental Engineering Professors; he now is a member of the editorial advisory board of the *Journal of the American Water Works Association* and vice-chair of the Drinking Water Committee of the Environmental Protection Agency's Science Advisory Board. He has been a member of several National Research Council committees, and he consults regularly for private industry and public agencies throughout the United States and Canada. He received his B.S. and M.S. degrees in civil engineering and his Ph.D. in water resources engineering from the University of Michigan.

Gunther F. Craun is president of Craun and Associates. He has had nearly a 30-year career in drinking water regulation and research. From 1965 until 1991, he was a commissioned officer in the U.S. Public Health Service. From 1971 until 1991, he was assigned to the Environmental Protection Agency's drinking water program and research and development office. He held positions as chief of the epidemiology branch and coordinator of environmental epidemiology for the

Health Effects Research Laboratory and assistant to the director of the Drinking Water Research Division in Cincinnati, Ohio. He has authored or coauthored several books in the drinking water field, including *Waterborne Diseases in the United States*, published by CRC Press, and *Methods for the Investigation and Prevention of Waterborne Disease Outbreaks*, published by the Environmental Protection Agency. He holds a B.S. in civil engineering and an M.S. in sanitary engineering from Virginia Polytechnic Institute and an M.P.H. and S.M. in epidemiology from Harvard University.

Stephen E. Himmell is vice president of engineering and Kankakee district manager for Consumers Illinois Water Company, the Illinois branch of Consumers Water Company that manages a larger number of water supply systems around the United States. In this position, he oversees management, operation, and budgeting of water and wastewater systems of the 18,000 customers of Kankakee district in Illinois and the engineering for the company's systems serving 60,000 customers in seven locations in Illinois. He has worked in the planning, design, management, and operations of water and wastewater systems for the past 22 years, the past 11 at Consumers. He recently was a member of a team that received the Build America Award of Merit from the Associated General Contractors for a redesign of a water system that saved the community $3 million. He holds M.S. and B.S. degrees in civil and sanitary engineering from the University of Missouri, Rolla, and is a certified professional engineer and Class A water works operator in the state of Illinois.

Carol R. James is the founder and principal of C. R. James and Associates, Consulting Engineers. She has more than 16 years of experience in conducting water quality and treatability studies and in water treatment plant pre-design. She has performed extensive alternative water treatment pilot-scale studies, water quality evaluations, regulatory compliance studies, and laboratory evaluations. She also has expertise in source water protection, having conducted watershed management projects and watershed sanitary surveys. She received a B.S. in civil engineering from the University of Virginia and an M.S. in environmental engineering from the University of North Carolina, Chapel Hill. She is registered as a civil engineer in the state of California.

Dennis D. Juranek is chief of the parasitology section for the Centers for Disease Control (CDC), where he has worked since 1970. His recent work has included conducting epidemiologic investigations of waterborne outbreaks of *Giardia* and *Cryptosporidium* around the country; he was involved in the CDC's investigation of the recent *Cryptosporidium* outbreak in Milwaukee. He has published widely on outbreaks of various parasitic diseases and methods for investigating outbreaks. In addition to his work at CDC, Dr. Juranek serves as a faculty member in the Division of Public Health at Emory University. He received a B.S. in biology from Colorado State University, a D.V.M. from Colorado State's College of Veterinary Medicine, and an M.Sc. in medical parasitology from the London School of Tropical Medicine and Hygiene.

Gary S. Logsdon is director of water treatment research for Black & Veatch. Previously, he served for more than 25 years with the U.S. Public Health Service and the Environmental Protection Agency. In his current position, he directs design of pilot drinking water treatment plant testing programs and works with water utilities to optimize their operations. He has a wide range of experience in water treatment technology development; he has conducted research on water filtration for removal of *Giardia* cysts, bacteria, and turbidity and on the modification of water quality for corrosion control. He holds B.S. and M.S. degrees in civil and sanitary engineering from the University of Missouri at Columbia and a D.Sc. from Washington University.

Frederick A. Marrocco is chief of the Drinking Water Management Division for the Pennsylvania Department of Environmental Resources. In this position, he is responsible for developing and administering Pennsylvania's safe drinking water program, including its special assistance program for small communities. Marrocco was formerly a member of the National Drinking Water Advisory Council and president of the Association of State Drinking Water Administrators. He holds an M.S. degree in environmental engineering from the University of North Carolina and a B.S. degree in civil engineering from Drexel University.

John M. Maxwell is vice president of Economic and Engineering Services, where he specializes in financial management and planning for small utility systems. Previously, he worked as a state regulatory engineer, assisting utilities with system improvements, operation and maintenance, and operator certification. He is a national instructor for the American Water Works Association in the areas of cost of water supply service and rate studies. He holds M.S. degrees in sanitary engineering and environmental science from Washington State University.

Daniel A. Okun, a member of the National Academy of Engineering and the Institute of Medicine, is a professor emeritus of environmental engineering at the University of North Carolina, Chapel Hill. He chaired the University of North Carolina's Department of Environmental Sciences and Engineering from 1955 until 1973. In addition to his academic responsibilities, he has advised cities, states, and governments worldwide on issues of water management and has served as a consultant for the World Bank, the Agency for International Development, and the World Health Organization. Dr. Okun chaired the Water Science and Technology Board from 1991 until 1994. He received his Sc.D. in sanitary engineering from Harvard University. He authored *Regionalization of Water Management: A Revolution in England and Wales,* published in London in 1977.

David R. Siburg is manager of the Kitsap Public Utility District, a countywide water supplier in Washington. The Kitsap district owns and operates 20 systems serving populations of from 10 to 5,000 people and provides contract service to more than 150 systems owned by others. Siburg received a B.A. in economics from Pacific Lutheran University and an M.A. in planning and public affairs from the University of Minnesota.

Velma M. Smith is director of domestic policy and director of the Groundwater Protection Project at Friends of the Earth. Previously, she served as legislative assistant on environmental and energy issues for Congressman Rick Boucher (D-VA), and prior to that she worked on farmland protection, land use, and water quality issues for the Piedmont Environmental Council in Warrenton, Virginia. While working for Congressman Boucher, Smith worked with small communities in western Virginia to help secure new potable drinking water supplies. She was involved in the 1986 reauthorization of the Safe Drinking Water Act for Friends of the Earth and has followed the implementation of the law since that time. Smith has frequently worked with the Rural Community Action Project to develop a better understanding among environmentalists of the water needs of small, low-income communities. In testimony before Congress, she has advocated regionalization of small drinking water systems. Smith served for 6 years as a member of the Virginia State Water Control Board, including one year as chair. She also served on the National Drinking Water Advisory Council. She holds a master's degree in environmental planning from the University of Virginia.

Amy K. Zander is an assistant professor of civil and environmental engineering at Clarkson University. Her research focuses on membrane technologies for removing contaminants from drinking water. Previously, she served as a water quality specialist for the Texas Water Commission. She received a B.S. in biology and M.S. and Ph.D. degrees in civil engineering from the University of Minnesota and is a registered professional engineer in the state of Minnesota.

Staff

Jacqueline A. MacDonald is a senior staff officer at the National Research Council's Water Science and Technology Board. She served as study director for the Committee on Small Water Supply Systems. Previously, she directed studies on alternatives for ground water cleanup and bioremediation of contaminated ground water and soil. She holds an M.S. in environmental science in civil engineering from the University of Illinois and a B.A., *magna cum laude*, in mathematics from Bryn Mawr College.

Etan Z. Gumerman is a research associate at the National Research Council's Water Science and Technology Board. He performed a variety of research duties for the Committee on Small Water Supply Systems. He has also served as project coordinator for a study of the economic valuation of ground water. Gumerman earned his M.S. in engineering and policy from Washington University in St. Louis and his B.A. in environmental science from the University of Pennsylvania.

Anita A. Hall is administrative secretary and senior project assistant at the National Research Council's Water Science and Technology Board. She served as project assistant for the Committee on Small Water Supply Systems, coordinating meeting arrangements for the committee and production of the committee's report.

David Dobbs is a freelance writer and editor specializing in environmental, building, health, and science issues. His book *The Northern Forest*, co-authored with Richard Ober, won the Vermont Book Publishers Association "Book of the Year" award in 1995. Dobbs has edited texts on sports physiology, construction, sailing, horticulture, and natural resource issues; he writes frequently for publications including *Popular Science, Forest Notes, The Boston Globe, Vermont, Vermont Life, Parenting,* and *Eating Well.* Originally from Texas, he received his B.A. degree from Oberlin College and now lives in Montpelier, Vermont.

Index